PSYCHEDELIC APES

Alex Boese holds a master's degree in the history of science from the University of California, San Diego. He is the bestselling author of *Elephants on Acid*, *Electrified Sheep* and *Hippo Eats Dwarf* as well as the creator and curator of the online Museum of Hoaxes (at hoaxes.org). He is also a daily contributor to WeirdUniverse.net. He lives near San Diego.

Alex Boese

Psychedelic Apes

**From parallel universes to atomic dinosaurs –
the weirdest theories of science and history**

PAN BOOKS

First published 2019 by Macmillan

This paperback edition first published 2021 by Pan Books
an imprint of Pan Macmillan
The Smithson, 6 Briset Street, London EC1M 5NR
*EU representative:*Macmillan Publishers Ireland Ltd, 1st Floor,
The Liffey Trust Centre, 117-126 Sheriff Street, Upper
Dublin 1, D01 YC43
Associated companies throughout the world
www.panmacmillan.com

ISBN 978-1-5098-6052-4

9 8 7 6 5 4

A CIP catalogue record for this book is available from the British Library.

Typeset by Jouve (UK), Milton Keynes
Printed and bound by CPI Group (UK) Ltd, Croydon, CR0 4YY

Visit **www.panmacmillan.com** to read more about all our books
and to buy them. You will also find features, author interviews and
news of any author events, and you can sign up for e-newsletters
so that you're always first to hear about our new releases.

CONTENTS

INTRODUCTION

This is a book about the weirdest, wackiest and most notorious against-the-mainstream theories of all time. In the following chapters, we'll explore curious questions such as, Are we living in a computer simulation? Do diseases come from outer space? What if planets occasionally explode? Is it possible the dinosaurs died in a nuclear war? Could humans be descended from aquatic apes? And was Jesus actually Julius Caesar?

Such notions may sound so outrageous that no one could possibly take them seriously, but they're not intended as jokes. Over the years, these odd speculations have been put forward in all seriousness by scholars who have argued that, no matter how much they might challenge the conventional wisdom, they could actually be true. Mainstream scientists, of course, strongly disagree. They insist that such ideas are nonsense. Some grow quite incensed that anyone would ever propose them in the first place, let alone believe them. Nevertheless, weird theories are a persistent presence in the history of science. They seem to sprout up constantly from the soil of intellectual culture, like strange, exotic growths.

As an exploration of unorthodox ideas, this book is part of a very old genre: the history of error. The purpose of this has traditionally been to describe supposedly foolish or incorrect beliefs in order to condemn them, holding them up as examples of flawed thinking to be avoided. That, however, isn't my intention here. Nor, on the other hand, do I want to defend or endorse these weird theories. My relationship to them is more complicated. I recognize that they make outrageous claims. I'm also quite willing to admit that most, perhaps even all of them, might be entirely wrong. And yet,

I'm not hostile to these theories. In fact, they fascinate me, and that's why I wrote this book.

On one level, I'm drawn to them because of a quirk of my personality. For as long as I can remember, I've had a preoccupation with oddities of history, especially ones involving outsiders and eccentrics. From this perspective, the appeal of these theories is obvious, because they're all the product of peculiar imaginations. Many of their creators were legendary misfits who ended up ostracized from the scientific community because of their insistent championing of aberrant notions.

I'm also intrigued by these theories because they offer a unique window into scientific culture, which is a fascinating subject in its own right. In particular, they reveal the tension between contrarianism and consensus-building that lies at its very heart.

Science is a unique form of knowledge in that it promotes scepticism about its own claims. It denies the notion of absolute certainty. It always admits the possibility of doubt, striving to put its claims to the test. For this reason, it places an enormous value on contrarianism, or being able to 'think different', as the famous Apple advertising slogan put it. Consider how scientific geniuses such as Copernicus, Darwin and Einstein are celebrated because they revolutionized our understanding of the natural world by seeing it in completely new ways.

But, simultaneously, science requires consensus building. It would be useless if researchers were forever disagreeing with one another, endlessly producing new rival explanations. At some point, they have to come together and accept that one interpretation of the evidence is more compelling than all the others. In other words, while science may shower its highest honours on those who can think differently, most scientists, most of the time, need to think the same way. As described by the historian Thomas Kuhn in his 1962 book *The Structure of Scientific Revolutions*, their jointly held interpretations, or paradigms, guide day-to-day research, shaping both the questions that get asked and the answers deemed legitimate.

So, both contrarians and consensus-builders play a necessary role in science, but, as we'll see, they often clash – though this may be putting it too lightly. It wouldn't be an exaggeration to say that they often end up despising each other outright.

The problem, as the consensus-builders see it, is that while contrarianism has its place, it can easily be taken too far. They argue that if a paradigm is supported by overwhelming evidence, then persisting in rejecting it, preferring to promote one's own radical theory in its place, can quickly degenerate into lunacy. It becomes tantamount to rejecting science itself.

The contrarians, on the other hand, stress that there are always different ways to interpret evidence and that the evidence might even be incomplete; perhaps a crucial piece of the puzzle is missing. They warn that rigid conformity can pose a far graver threat to science, because it blinds researchers to possible new interpretations.

I find myself sympathetic to both sides in this debate. I accept that, realistically, the conventional scientific wisdom is almost certainly right. Scientists, after all, are highly trained to evaluate evidence. If it has persuaded most of them to favour one interpretation, it's probably because that genuinely is the best one. But I have enough of a contrarian in me to find myself happy that the iconoclasts are out there asking awkward questions, stirring up the pot – even if, at times, they may come across as totally nuts. On occasion – maybe not often, but every now and then – the wild, unorthodox theory that seems to defy common sense does end up being vindicated.

Which leads to the main reason for my fascination with weird theories: sheer curiosity! When someone comes up with a truly outrageous idea that completely flies in the face of orthodox opinion, I can't help it – I want to know what their argument is, and part of me wonders, Is it possible they could be right? Is their alternative point of view simply crazy, or could it be genius?

That's the fun of these theories. They offer up the thrill of unbridled speculation. They tackle some of the greatest questions in science – about creation, the nature of the universe, the origin of life and our species, the emergence of consciousness and the rise of civilization – while advancing seemingly absurd answers to them. But are the answers really that crazy, given how many unknowns surround all these topics? That's the element of uncertainty that serves as their hook, giving these theories their power to win converts.

By exploring these heretical ideas, you can venture briefly into intellectually off-limit areas and you can find out if any of them can (possibly) seduce you to their side. Perhaps they'll cast doubt on subjects that you thought were entirely settled, or that didn't even seem to be problems at all.

It's my intention to give these theories a chance to persuade you. Therefore, I've tried to provide a fair reconstruction of what the arguments for them are, even if that risks making me seem overly sympathetic towards them at times. Although, in every case, I've also made sure to explain why these theories are rejected by the appropriate experts. What I won't do is tell you whether I think any of them are right or wrong. I'll let you decide their worth for yourself.

And if you do end up concluding that the conventional wisdom is correct and that these theories really are crazy, I'd maintain that the time spent considering them still won't have been wasted. The classicist Mary Beard once argued, while discussing the theory that Homer was a woman (which we'll examine in the final section of this book), that it's possible for an idea to be 'usefully wrong'. I fully agree with her. I believe that a provocative idea, even if it's absolutely mistaken, can jolt us out of our intellectual rut. The exercise of stepping outside of what's familiar can shake up our thinking and provoke us to question our assumptions, revealing that it might be possible to look at what we took for granted in an entirely new way.

And, on a somewhat more practical level, these weird theories do offer an offbeat way to learn quite a bit about standard science, because they constantly engage with the dominant scientific schools of thought, even while disputing them. It's like taking a funhouse-mirror tour of the history of science. The frame of reference may be highly unconventional, but you will nevertheless be exposed to a lot of what is considered 'correct' science mixed in with the incorrect. Hopefully, some of these theories might even inspire you to hunt down more information about the topics or to explore a discipline further.

The genre of weird theories is vast and sprawling. It would have taken multiple volumes to explore it completely. To make things more manageable, I applied some filters.

First and foremost, I eliminated any theory that I didn't believe to be genuinely contrarian in spirit. Right off the bat, this removed from consideration two ideas that are widely associated with resistance to scientific orthodoxy: creationism and climate-change denial. Many of the theories we'll examine in this book have been accused by critics of being similar to these two. But I think the comparison is misguided.

Creationism is inspired by a rigid commitment to an ancient religious orthodoxy. That's the opposite of contrarianism. Climate-change deniers similarly represent a powerful interest group: the fossil-fuel industry and all its allies.

A true scientific contrarian, in my mind, hasn't simply embraced an alternative, pre-existing orthodoxy. They're not the attack dogs of some well-entrenched faction that perceives itself to be threatened by the scientific consensus. Instead, they're genuine oddballs who have carved out their own peculiar niche. Whatever else one may say about them, there's no cynicism or ulterior motives lurking behind their stance. They're honestly passionate about seeking the truth, as they perceive it. And while they're definitely opposed to specific scientific ideas, they're not anti-science. Just the opposite. They view themselves as the legitimate upholders of scientific values, fighting against the rise of groupthink.

As a corollary to this, I believe that ignorance is not the same as contrarianism. A true contrarian has to know the science they're rejecting. There's an entire genre of crackpot science in which people dream up elaborate theories, apparently never having read any of the relevant scientific literature on the topic. I view that as a separate phenomenon – interesting in its own way, but not what I want to examine here. All the theorists we'll look at have, I feel, made an effort to understand the paradigms they reject. In fact, quite a few of them were highly respected figures in their disciplines before, to the bewilderment of their colleagues, seeming to go completely off the rails.

Even with this filter, our topic remains huge. So I further narrowed down the focus to the historical sciences: cosmology, geology, evolutionary biology, palaeontology, anthropology and the social sciences, such as archaeology and history. This is in contrast to the experimental sciences, such as physics and chemistry. (I let two physics theories slip through because they're both relevant to cosmology.)

I chose this selection because, for my money, the historical disciplines produce the best (i.e. most outrageous) weird theories. They're home to some of the all-time classics of the genre. They're inherently more speculative than the experimental sciences and so theorists can really let their imaginations fly.

Concentrating on the historical sciences has also allowed me to add some structure to the book. While each chapter stands on its own, and you can read them in any order you wish, if you do choose to read the book from front to back, the topics will proceed in a rough thematic sequence. We'll start with the widest-scale view possible, the entire universe, and then we'll progressively zoom in closer: on to the solar system, then the Earth, the origin of life, the evolution of humans and, finally, the emergence of civilization. In this way, I've designed the book to offer a kind of alternative history of the cosmos, from its creation right up to the dawn of the modern era.

For the sake of variety, I've also thrown in along the way a few examples of weird-became-true theories: concepts that were initially rejected by the scientific community, but were eventually accepted as correct (or, at least, plausible). I did this to show that it is indeed possible for theories to make that journey from being outcast to being accepted.

Finally, let me add a note about terminology. The word 'theory' has a specific meaning in science. It's defined as an explanation that's strongly supported by evidence and generally accepted by the scientific community. This contrasts with a hypothesis, which is more like an educated guess based on limited evidence.

A problem arises because, in popular usage, a 'theory' means any kind of supposition or idea. The term is interchangeable with a hypothesis or speculation. This creates a source of tension because some scientists can be very particular about the usage of these words. In fact, they've been waging a campaign for over two centuries to try to get people to stop referring to hypotheses as theories. They worry that, if the public perceives a theory to be any old idea or conjecture, then they won't attach sufficient weight or importance to it. They may end up thinking that the theory of evolution is just some half-baked notion that Charles Darwin dreamed up while sitting on the toilet.

I'm afraid I'm going to draw down the ire of the linguistic sticklers, because I've opted to use 'theory' in its lay rather than its scientific meaning. My excuse is that this is a book for a broader audience, so I thought it fair to use the word as it's understood in general usage. I've tried to stick with whatever term was most widely used to describe each idea. If most people called it a theory (whether or not it was accurate to do so), so have I. Likewise, if most people have referred to a certain concept as a hypothesis, I do too.

Which is all a long-winded way of saying that, just because I may refer to some of the ideas in the following pages as theories, it doesn't mean that technically any of them are. A few of them do actually come close, and you'll find a range of plausibility in the following pages. I suspect you may even end up agreeing with some of the claims! Others, however, don't even really pass muster as hypotheses. Mere conjectures might be more accurate. Proceed with appropriate caution.

CHAPTER ONE

Cosmological Conundrums

If you travel out past the skyscrapers and city lights, through the suburbs and into the open countryside beyond, be sure to look up at the sky at night. You'll see the universe stretched out above you in all its vastness. It's an awe-inspiring sight, and it may lead you to wonder – what exactly is it and where did it all come from? How did the stars and galaxies get there? Just how can we make sense of this immensity?

People have been pondering such questions since prehistoric times, when they found answers in mythology. Ancient Chinese myth taught that the universe formed when a giant named Pan Gu awoke inside an egg and shattered it to get out. In Lenape and Iroquois creation tales, a sea turtle carries the entire world on its back, and, according to the Babylonian epic *Enûma Eliš*, the storm god Marduk fashioned the heavens and Earth out of the slain body of a massive sea dragon, Tiamat.

Nowadays, it's the discipline of cosmology that tackles questions about the universe as a whole. By its very nature, it deals with concepts abstract and strange. Black holes, higher dimensions and virtual particles flickering in and out of existence in the vacuum of space, to name a few. As a result, cosmologists have a very high tolerance for odd ideas. The theories they regard as orthodox can be mind bending. As for the unorthodox theories we'll examine in this section . . . let's just say that they call into question some of the most basic aspects of what we consider to be reality.

What if the Big Bang never happened?

How did the universe come into existence? Mainstream science tells us that it burst forth explosively from a massively hot, dense state, approximately 13.8 billion years ago. As this cosmic fireball expanded, it cooled, forming first into atoms, and then into stars and galaxies, and finally a small part of it transformed into the Earth and all its inhabitants, including you and me.

This is the Big Bang theory, which first took shape back in the 1920s, though it didn't achieve widespread acceptance right away. For several decades during the mid-twentieth century, it faced a serious challenge from a theory called the steady-state model. This alternative cosmology, proposed in the late 1940s, made the case for a radically different creation scenario in which the universe had no beginning and the Big Bang never happened. In fact, steady-state advocates argued that a fundamental principle of science prohibited the Big Bang from ever having happened.

They didn't propose, however, that this lack of a beginning meant that creation itself had never occurred. Just the opposite. They maintained – and this was the oddest part of the theory – that creation was going on all the time. They envisioned bits of matter continuously materializing out of the void in the far reaches of space. Exactly what form this new matter took wasn't clear. The authors of the theory speculated that it was probably stray hydrogen atoms that popped into existence, though one of them whimsically suggested it might even be spontaneously emerging cakes of soap, but they contended that, whatever form the new matter took, the universe endlessly renewed itself by means of this process – a bit like, if it were possible, a person achieving immortality through a perpetual infusion of young, healthy cells.

Crack open any science textbook and it's not going to say anything about stray atoms or cakes of soap popping into existence out of nothing. Modern science absolutely doesn't believe such a phenomenon is possible. It would be more like magic. So, how did the authors of the steady-state theory convince themselves that this idea of continuous creation might be a reality? To understand this, let's back up a bit and first examine the genesis of the Big Bang theory, because the one led to the other.

It was observational evidence that provided the inspiration for the Big Bang theory. During the 1920s, the astronomer Edwin Hubble, using the large new telescope that had recently been installed at the Mount Wilson Observatory in California, discovered that almost all the galaxies in the visible universe were rapidly receding from one another, as if fleeing outwards. This led him to conclude that the universe must be expanding.

This discovery, in turn, quickly led the Belgian physicist (and Roman Catholic priest) Georges Lemaître to reason that, if the universe is getting bigger, it must have been smaller in the past. Much smaller. If one were to reverse time back far enough, he surmised, one would eventually arrive at an initial moment when all the material in the universe was compressed together into a single small mass, which he called the 'primeval atom'. Everything in existence, he argued, must have come from this one source. This logic was compelling enough to rapidly establish the Big Bang as the leading scientific theory about the origin of the cosmos.

The steady-state theory developed subsequently, but its inspiration came from more abstract, philosophical concerns. It was the brainchild of three Cambridge researchers: Hermann Bondi,

Thomas Gold and Fred Hoyle. Bondi and Gold were both Austrian émigrés who had fled Nazi Germany, while Hoyle was a native of Yorkshire, in England. They met when the British army put their scientific talents to work researching radar during World War II, and they continued their friendship as young professors following the war.

All three agreed that Hubble's discovery of the expansion of the universe was important, but they felt that Lemaître's conclusion had to be incorrect because, they believed, it contradicted a fundamental principle of science – this being that the laws of nature are universal and apply uniformly everywhere *and at all times*. They insisted that this was an absolute concept which couldn't be compromised. They warned that, if you started messing with it – if, for instance, you decided that the law of gravity worked on Tuesdays, but maybe not on Wednesdays – then the entire structure of science would collapse. Knowledge would become impossible.

This principle, as such, wasn't controversial. They were right that it was a fundamental part of scientific belief. But, when they rigidly applied it to the question of the origin of the universe, it led them to the startling conclusion that the creation of matter and energy couldn't have been a one-time event, as Lemaître assumed, because, if creation had been possible once (which it evidently had been, because we exist), then it must always be possible and always will be. Creation had to be an ongoing process. If the laws of nature are constant throughout time, how could it be otherwise?

They criticized Lemaître's theory as being profoundly unscientific because it violated this principle, leaving creation unexplained as a mysterious, one-time event at the beginning of time. Bondi scolded, 'To push the entire question of creation into the past is to restrict science to a discussion of what happened after creation while forbidding it to examine creation itself.'

This, in a nutshell, was the dispute between the two cosmological models. Big Bang advocates appealed to observational evidence that suggested a one-time creation event, whereas steady-state proponents, in response, invoked a philosophical principle to insist that creation had to be ongoing and continuous.

Bondi, Gold and Hoyle conceded that the idea of matter being continuously created would strike many as strange. After all, there were other scientific principles that needed to be considered, such as the law of conservation of energy. This states that energy can neither be created nor destroyed; it can only change form. Therefore, matter, being a form of energy, shouldn't ever pop into existence out of nothing.

Just as importantly, there wasn't a shred of observational evidence to support the claim of continuous creation. Scientists had never witnessed anything like such a phenomenon. The physicist Herbert Dingle angrily compared the concept to the alchemical belief that lead can be changed into gold by means of occult magic.

Nevertheless, the Cambridge trio still insisted that it was more reasonable to assume creation was an ongoing process rather than a one-time event. To support this contention, they carefully worked out the details of a cosmological model based upon never-ending creation to demonstrate how it could plausibly work. This led to their steady-state model.

The story goes that the three researchers initially came up with the grand vision of their alternative cosmology after watching a 1945 horror movie, titled *Dead of Night*, about a man trapped in a recurring nightmare. The movie ended with the man waking up once again at the beginning of his dream, and this looping narrative structure made the researchers think of a universe with no beginning or end. That connection probably wouldn't seem obvious to most people, but the bottle of rum the three were sharing after the movie, as they sat in Bondi's apartment, evidently helped the analogy make sense to them.

Whereas the Big Bang universe began with a violent, explosive event and then underwent dramatic change over time, their steady-state universe was all calmness and serenity. It offered the reassuring vision that, on a sufficiently large scale, the cosmos always has and always will appear the same.

In his 1950 textbook on cosmology, Bondi explained that the term 'steady state' was meant to evoke this idea of a universe that always maintains the same large-scale appearance. He compared it to a river. The water in a river constantly changes as it flows downstream, but the overall aspect of the river remains the same from one day to the next. The river maintains a steady rate of flow. Likewise, the steady-state universe would change on a small scale all the time, but its overall aspect remained forever the same.

Continuous creation was the key to maintaining this stability. If whatever existed was all that ever would, and no new matter ever came into being, then the universe would eventually fade into a cold death as stars burned through all the available fuel and went dark. But continuous creation provided a never-ending supply of fuel, allowing new stars to form even as old ones burned out.

By their calculations, it didn't even require a lot of matter-creation to keep the universe running. The amount was so small that a person would never be able to see it happening, nor could any known experiment detect it. As Hoyle, who was known for his homespun explanations, put it: 'In a volume equal to a one-pint milk bottle about one atom is created in a thousand million years.' It was also Hoyle who suggested, somewhat tongue-in-cheek, that the matter-creation might take the form of cakes of soap.

As for the law of conservation of energy that forbids the creation of new matter, Hoyle argued that it was actually possible for continuous creation to happen without violating this law. The trick that allowed it was negative energy. Hoyle hypothesized the existence of a universe-wide field of negative energy, which he called a 'creation field' or C-field. Any disturbance of this creation field, he said, would cause it to increase in size, which then triggered the creation of an equivalent amount of positive energy (aka matter). The simultaneous creation of positive and negative energy cancelled each other out, leaving the total sum of energy in the universe constant.

Critics dismissed this as a mathematical trick, but Hoyle responded that it nevertheless worked, because the law of conservation of energy only required that the *total* amount of energy remained the same. It didn't matter how much positive and negative energy came into existence. As long as the sum total balanced out, the conservation law wasn't violated.

The creation of matter also served a second purpose. As it formed, the creation field grew, and, because its energy was negative, it had an antigravity effect, causing the universe to expand. This fitted in with Hubble's earlier observations. The expansion itself then acted as a kind of cosmic trash sweep. Old stars and galaxies were pushed outwards, past the edge of the observable universe, allowing new ones to take their place. In this way, all parts of the steady-state system worked together like a smoothly running piece of machinery, on and on for eternity, with no beginning and no end.

The Cambridge trio published the details of their new cosmology in 1948. It appeared as two articles: one authored by Bondi and Gold, and the second by Hoyle alone. They then set about promoting the theory. It's one of the small quirks of history that, in the course of doing so, Hoyle accidentally gave the Big Bang its name. Until that time, Lemaître's theory had usually been referred to as the 'evolutionary cosmology' model, but during a BBC radio lecture in 1949, Hoyle described it, somewhat dismissively, as the idea of matter being created 'in one Big Bang at a particular time in the remote past', and the phrase stuck.

For a while, the steady-state theory gained a modestly large following, especially among British researchers. Historians of science have noted that the theory, by keeping the entire cosmos nicely calm, steady and unchanging, seemed to appeal to the British love of stability and tradition. Hoyle was also a powerful and influential champion. In the mid-1950s, he led a team that worked out the physics of stellar nucleosynthesis – how elements such as carbon and iron are forged out of hydrogen and helium inside suns. This discovery is considered to be one of the greatest achievements in astrophysics of the twentieth century.

What eventually did the theory in, however, was observational evidence, which had always been the stronger suit of the Big Bang theory. As astronomers continued to explore the universe, they found that it simply didn't look the way the steady-state model predicted it should.

During the 1950s, astronomers had begun using the new technology of radio telescopes to peer deep into the most distant, and therefore oldest, parts of the cosmos. What they found was that the galaxies in those regions were more densely packed together than they were in younger parts of the cosmos. This directly contradicted the steady-state prediction that the universe should always have had the same appearance (and therefore density) as it does now.

The real knockout blow, however, came in 1965 with the discovery of the cosmic microwave background. This is a faint whisper of electromagnetic radiation filling every corner of the universe. Big Bang theorists had predicted that exactly such a phenomenon should exist, left behind as a radiant afterglow of the extremely hot initial conditions of the early universe. Steady-state advocates, on the other hand, were caught flat-footed. They didn't have a ready explanation for why this cosmic background radiation was there.

In the opinion of most scientists, these pieces of observational evidence, taken together, tipped the balance decisively in favour of the Big Bang theory. They clearly indicated that the universe must have had a beginning. As a result, the steady-state theory rapidly lost support and, by the 1970s, the Big Bang had gained acceptance as the standard model of cosmology.

Not to be beaten, in the 1990s Hoyle tried to cobble together a comeback for the steady-state theory. He partnered with the astrophysicists Jayant Narlikar and Geoffrey Burbidge, and together they devised what they called the quasi-steady-state cosmology.

In this new version, they conceded that observational evidence did indicate some kind of big, cosmic-scale event had occurred approximately fourteen billion years ago, but they argued that this event didn't necessarily need to be the origin of the universe. They proposed, instead, that the universe went through endless fifty-billion-year cycles of contraction and expansion. They didn't imagine that it contracted all the way down to the size of Lemaître's primeval atom. During its most recent contraction, they said, it had remained large enough for entire galaxies to remain intact. This differentiated their idea from some models of the Big Bang that envision the universe going through cycles of collapse and rebirth. But it shrank enough, they claimed, that what appeared to be the Big Bang was actually the last contraction phase ending and the present expansion phase beginning. This allowed all the observations cited as evidence for the Big Bang to be reinterpreted within the framework of continuous creation.

To most astronomers, this new model seemed little more than a desperate attempt to save a failed theory, and they basically ignored it. It certainly did nothing to put a dent in the popularity of the Big Bang theory. Sadly, Hoyle died in 2001, and with his death the steady-state model lost its most vocal and prominent advocate.

Given this history, it would seem easy to dismiss the steady-state theory as an ambitious but misguided attempt to found a cosmological model upon a philosophical principle rather than observational evidence. Certainly, it must have been doomed to failure! But the thing is, the theory actually raised a legitimate question: how *did* the creation of matter and energy occur? How did something emerge out of nothing? If bits of matter randomly popping into existence throughout deep space is the wrong answer, then what is the right answer? What is the Big Bang explanation for the phenomenon?

Up until the 1970s, the prevailing school of thought among Big Bang advocates was simply to treat creation as an off-limits subject. The evidence strongly indicated that a single creation event had occurred, but there was no clue as to why it had happened or what caused it, so it seemed pointless to speculate about it. But when the Big Bang became enshrined as orthodoxy, ignoring the question of creation began to feel unsatisfactory. As the historian of science John Hands has said, it became like the elephant in the room of modern cosmology. A lot of scientists felt that some kind of explanation was necessary.

Of course, there was always the God solution. Perhaps a divine being had caused creation with a snap of its fingers. It didn't escape the notice of opponents of the Big Bang that the theory seemed peculiarly compatible with this explanation. Nor did it escape the notice of Pope Pius XII, who, in 1951, praised the theory for offering scientific proof that the universe had a creator. Over the years, many of the staunchest Big Bang critics have been atheists who have accused it of being little more than a device to surreptitiously sneak theology into science. Lemaître, they point out, was both a physicist and a priest.

Big Bang advocates vehemently denied this charge, noting that many of them were atheists too! Anyway, they didn't want a religious solution to the mystery of creation; they wanted a proper scientific one. But the problem they encountered, as they pondered how to explain creation within the context of the Big Bang, was that the critique by steady-state proponents was actually right. It *was* very odd that creation would only have happened once. The fundamental forces of nature, such as gravity and electromagnetism, are all ongoing. They don't turn on and off. So, why would some force have allowed creation to happen once, but then forbidden it from ever occurring again?

This was the puzzle, and logic led inescapably to one possible, though paradoxical, answer: if creation can't occur in our universe, then it must have happened elsewhere. As bizarre as it might sound, there must be more to the cosmos than our universe, and creation must somehow have occurred (and possibly still is occurring) out of sight, in that other region.

This explanation required a redefinition of the term 'universe'. Traditionally, the word had referred to absolutely everything in existence, but now it was given a more limited meaning. The term 'cosmos' continued to refer to everything there was – the whole shebang. The universe, however, was redefined to mean everything created by the Big Bang. This implied that vast regions that lay outside the creation event that formed our particular universe might exist. There might, in fact, be many more universes than just the one we inhabit.

Since the 1970s, some variety of this answer has been adopted by most mainstream cosmologists. This represents a significant modification of Lemaître's original Big Bang model, which claimed to describe the creation of absolutely everything. The new Big Bang only describes the creation of our particular universe out of a pre-existing something. There's no consensus, however, about what this something might have been. Perhaps it was a quantum vacuum in which random fluctuations of energy occasionally produced new universes. Perhaps it was a five-dimensional hyperspace inhabited by floating membranes of energy that spawned new universes every time they collided. Or perhaps (and this is currently the most popular belief) it was a 'multiverse' filled by a rapidly inflating field of negative energy out of which new universes constantly formed, like drops of water condensing out of steam.

But consider the significance of these speculations. They suggest that, if the Big Bang did happen, the cosmos as a whole must be a very strange, unfamiliar place – both vastly larger than our universe and profoundly different in character from it as well. By contrast, if the Big Bang never happened, as the steady-state theory envisioned, then the universe as we see it is pretty much the way the entire cosmos actually is, everywhere and always.

Seen from this perspective, the steady state is revealed to have been a deeply conservative theory. It accepted a small bit of weirdness (continuous creation occurring within our universe) in order to achieve the pay-off of a greater overall normalcy, preserving our universe as the entirety of the cosmos. The Big Bang, on the other hand, rejected the weirdness of continuous creation, but as a result its proponents ended up exporting creation outside of our universe. They reimagined our universe as a tiny part of a far greater whole – a kind of bubble universe, floating in an infinite alien landscape, surrounded by other bubble-verses.

This is the irony of the steady-state model. With its notion of matter promiscuously popping into existence everywhere, it's come to be considered an unorthodox, weird theory, but its model of the cosmos is arguably far less radical than the ones dreamed up by proponents of the current Big Bang theory. So, which is really the weirder theory? Perhaps the reality is that there is no non-weird way of addressing the question of creation. All efforts to solve this mystery lead to some very odd implications.

Weird became true: radio astronomy

During the second half of the twentieth century, radio telescopes revolutionized astronomy by opening an entirely new window onto the universe. They allowed researchers to discover objects in the cosmos, the existence of which had never previously been suspected, such as highly energetic galaxies called quasars and fast-spinning neutron stars called pulsars. So, surely astronomers greeted the introduction of radio astronomy with open arms? Not quite. In fact, the initial reaction was more along the lines of a collective shrug of their shoulders.

The first hint that such a thing was possible didn't come from the astronomical community at all. It came from Bell Telephone Laboratories, the research division of AT&T, which in the early 1930s had become interested in the possibility of using radio for transatlantic phone calls. During test calls, however, the connection kept getting interrupted by static coming from an unknown origin. The company assigned a young engineer, twenty-six-year-old Karl Jansky, to track down the source of the interference.

To do this, Jansky built a hundred-foot rotating radio antenna in a field on an abandoned potato farm near the headquarters of Bell Labs, in Holmdel, New Jersey. His colleagues nicknamed the device 'Jansky's merry-go-round'. After two years of investigation, he determined that local and distant thunderstorms were one cause of interference, but there was another source he just couldn't identify. It was a static-filled radio signal that peaked in intensity approximately every twenty-four hours.

By rotating his antenna, Jansky could pinpoint where the signal was coming from, and, to his surprise, this initially indicated it was coming from the sun. That alone was significant, because no one before had considered the possibility of radio waves coming from space, but as Jansky continued to track the signal, he realized it wasn't actually coming from the sun. Over the course of a year, the signal slowly travelled across the sky: it began in alignment with the sun; after six months, it was on the opposite side of the sky; and at the end of the year it had returned to solar alignment.

Jansky had no background in astronomy, but he knew enough to realize that this curious movement meant that the signal occupied a fixed position in the sky. It was the annual passage of the Earth around the sun that was making it appear as if the signal was moving. This, in turn, meant that the signal had to be coming from a source outside the solar system, such as a star, because no object inside the solar system maintains a fixed celestial position. After consulting star maps, he figured out that the signal seemed to be coming from the centre of the Milky Way, the galaxy that contains our solar system.

Jansky's discovery generated excited headlines in the press, as reporters were eager to know if he had picked up a broadcast from an alien civilization. He assured them it appeared to be caused by a natural phenomenon, because the signal was absolutely continuous and pure static. It sounded 'like bacon frying in a pan'. Nevertheless, the press found a way to sensationalize the news, further speculating that the radio beam might be a source of unlimited electrical power streaming from the centre of the galaxy. (No such luck, unfortunately.)

Professional astronomers, on the other hand, seemed unmoved by Jansky's find. They treated it as little more than a random curiosity. The problem was that your typical astronomer in the 1930s knew almost nothing about radio engineering. They peered through optical telescopes. They didn't tinker around with radios.

Jansky's anomalous find lay outside their area of expertise. Plus, the conventional wisdom was that stars produced light and nothing else. Some suggested that the true source of the signal might be stellar radiation striking the Earth's atmosphere, but most simply filed his report away and ignored it.

And that was the anticlimactic birth of radio astronomy. Jansky tried to convince Bell Labs to fund more research into the mysterious 'star noise' he had found, but the company saw no profit potential in that. Jansky's boss told him to move on to other projects, which he did.

Luckily for astronomy, while the professionals may not have been particularly excited about Jansky's discovery, there was one young man who was. This was twenty-one-year-old Grote Reber, an amateur radio enthusiast with a degree in electrical engineering, who lived in the Chicago area. Like Jansky, he had no background in astronomy, but he thought the star noise was the most amazing thing he had ever heard of and he decided to get to the bottom of the mystery.

At first, Reber tried to satisfy his curiosity by contacting the experts. He wrote to Jansky, seeking a job as his assistant, but Jansky informed him that Bell Labs had cut funding for the project. Then Reber checked to see if any astronomers were working on the problem. Harvard Observatory politely responded that Jansky's find was interesting, but they had more pressing research priorities to pursue. Gerard Kuiper, professor of astronomy at the University of Chicago, was more dismissive. He assured Reber that Jansky's discoveries were 'at best a mistake, and at worst a hoax.'

However, Reber had a contrarian streak that would stay with him his entire life. One of his biographers later wrote, 'Reber paid no attention to establishment science, except to express his disdain.' If the experts weren't going to pursue the puzzle of the 'star noise', Reber decided he would do it on his own.

In 1937, he began building the world's first radio telescope in an empty lot next to his mother's house, in Wheaton, Illinois, where he was living. Unlike Jansky's antenna, this was a proper radio telescope, featuring a 32-foot parabolic dish to focus the radio waves. The telescope rested on a large scaffold structure, which his mother proceeded to use to hang out her laundry, and the neighbourhood kids played on as a jungle gym.

Reber completed the telescope in 1938, and then set about producing the first ever map of the radio sky. He had to do this late at night, in part because he was working a day job at an electronics company in Chicago, but also because there were fewer cars on the road then; the static produced by their engines interfered with his sensitive receiver.

The following year, he sent details of his work to astronomical journals, but, like Jansky before him, he encountered a lack of interest, if not outright scepticism. He was, after all, mostly self-taught and lacked any academic affiliation. Editors at the journals weren't sure if he was for real or a random nutcase. Eventually, the editor of the *Astrophysical Journal* decided that he should give this young man a closer look, in case there was something to his claims. So he sent a team of astronomers out to Wheaton to examine the radio telescope.

They walked around the device in astonishment, poked and prodded it a bit, and finally reported back that it 'looked genuine'. The journal published a short piece by Reber in 1940 – the first publication about radio astronomy to appear in an astronomical journal. In this way, thanks to Reber's persistence, astronomers finally became aware of how radio waves could help them in their exploration of the universe.

Even so, it wasn't until after World War II that radio astronomy became fully established as a discipline, aided significantly by military interest in the development of radar technology. Old attitudes of indifference, however, continued to linger among astronomers for a number of years. According to one story, during an academic conference in the early 1950s, a highly regarded astrophysicist introduced the presentation of a young radio astronomer with the remark, 'Well, next one is a paper on radio astronomy, whatever that may be.'

Nowadays, of course, no astronomer would make a comment like that, as radio telescopes have become one of the most important tools at their disposal. The largest array of them ever, the Square Kilometre Array, is set to be built in Australia and South Africa and will come online in 2024. Budgeted at an initial cost of over $700 million, it's anticipated that it will be able to conduct the most accurate tests to date of Einstein's general theory of relativity, make fundamental discoveries about the nature of the cosmos and possibly even detect the presence of extraterrestrial life, if any is out there.

What if our universe is actually a computer simulation?

Scientists spend countless hours trying to understand how nature really works, but what if all their research is just wasted effort because everything in the universe is no more than a grand illusion? What if we're not actually flesh-and-blood creatures living on planet Earth, but instead we're bits of electronic data shuttling around inside a processor on a silicon chip? What if our consciousness and everything we sense and experience has been generated inside a computer that may be sitting on someone's desk in the 'real' world?

In 2003, the Oxford philosopher Nick Bostrom published an article in which he made the case that this unsettling notion isn't just pie-in-the-sky speculation. He insisted that there are logical reasons to believe it may be true. All of us, and the entire observable universe, may be computer simulations.

The idea that the world around us is a mere fabrication has been rattling around in philosophical circles for a long time. You can find references to it in ancient writings. Plato wrote that we're all like prisoners in a cave, staring at shadows on the wall and believing those shadows to be the real world, ignorant of the richer reality outside. Similar sentiments appeared in the earliest texts of Hinduism and Buddhism. But throughout this lengthy tradition of suspecting that our senses may be deceiving us, exactly how this deception occurs has always remained hazy. Perhaps a divine being had designed the world to be that way.

The invention of the computer in the twentieth century added a new twist to these doubts, because suddenly it became possible to imagine a physical means by which a fake reality could be created. The rapid advancement of technology made it seem increasingly plausible that researchers would one day be able to build an artificial, non-biological intelligence that possessed a consciousness equivalent to that of a human. Essentially, a brain living on a silicon chip. And if they could create such a being, then presumably they would also be able to control its sensory input. They could fashion a virtual computer-generated environment inside of which it would live. The consciousness would inhabit a simulated world, but it would have no way of knowing this.

This possibility posed a paradox: if it's technologically feasible for a conscious being to live in a simulated world without being aware of it, how can we know that this isn't our own situation? How can we be sure we're not one of those artificial brains inside a computer processor?

A few researchers have tried to answer this puzzle by figuring out ways of detecting the difference, presuming that we can science our way out of this paradox. They believe a simulated universe would inevitably contain telltale flaws if examined closely enough. It might appear blurry on the smallest scales, in the same way that digital images become pixelated if you zoom in too far. Or perhaps it would contain incriminating glitches and bugs in the programming that would give away the deception.

But is this line of thinking really foolproof? For a start, these researchers have assumed we know what a real world should look like. If we've actually been living inside a simulation for our entire lives, we wouldn't know this. We'd have no 'real' standard to judge our fake world against.

Their argument also assumes it's possible to outsmart the programmers of the simulation. But surely they would hold all the cards in this game, possessing many ways of hiding the truth from us, if they wanted to. If we ever did stumble upon indisputable evidence of fakery, they could simply rewind the program and edit out our discovery. Not to mention, there's no way to know if we've even been given free will to examine our world. For all we know, we might be performing scripted actions, naively believing that the decisions we make are our own. Are you sure you really wanted that second cup of coffee this morning, or were you just following commands?

In other words, the simulation paradox seems to be inescapable. If it's possible for an artificial intelligence to exist inside a computer-generated environment, then we can never escape the lingering uncertainty that our world might be a simulation.

Here's where Bostrom's argument enters the picture. He took it for granted that there was indeed no scientific method of determining the reality of our universe. Instead, he decided we could use probability analysis to figure out which scenario was more likely: whether we're living in a simulation or in the real world.

The bad news, he believed, was that, if we view the problem like this, treating it as a matter of statistics and odds, then we're naturally led to the conclusion that there's a decent chance we're in a simulation.

His reasoning was that, if you were to conduct a census of all the sentient beings that have ever existed or will ever exist in the history of the universe, then it's plausible you'd discover that the vast majority of them are sim-beings, possibly by a factor of ninety-nine to one or even greater. So it makes sense to conclude that we're most likely among the larger group: the sims.

The reason simulated intelligences probably vastly outnumber non-simulated ones is because there's only one real world, but there can potentially be many artificial ones. An advanced civilization with enormous amounts of computing power at its disposal could conceivably create thousands or even millions of fake 'worlds' filled with sim-beings.

In fact, we're already busy building ever more elaborate virtual worlds with existing technology. Computer games that involve simulated environments, such as *World of Warcraft*, *Second Life* and *SimCity* are hugely popular.

As technology continues to advance, it seems safe to assume that such games will continue to grow more sophisticated, becoming increasingly lifelike, until finally our descendants might populate them with full-fledged artificial intelligences. Bostrom believed that advanced civilizations might create artificial worlds not only for entertainment, but also for scientific research, as a way to study their ancestors and their own evolution.

Add to this the possibility of simulations being created within simulations, leading to even more artificial beings, and there might exist countless layers, nested within each other like Russian dolls, multiplying the number of simulations exponentially. But there will always remain just one real world, so the sim-beings would have a distinct numerical advantage over flesh-and-blood creatures.

So, surely that's it. We really are living in a computer simulation. But Bostrom cautioned that this line of argument has its limits. It rests on the assumption that advanced civilizations will both be able to develop simulations and will actually want to. After all, it's possible to imagine future scenarios in which simulations never get built. For all we know, advanced civilizations might destroy themselves or be wiped out before being able to reach the stage of development at which they could build truly convincing artificial worlds. Or perhaps we're underestimating the technological challenge and it may be impossible, anyway. Or they may be deemed totally unethical and banned as illegal.

All these possibilities are plausible, and they lower the probability that sim-beings outnumber real ones. Bostrom figures that, when all these different factors are considered, there's about a 20 per cent chance we're living in a simulation. That number is a lot better than a 99 per cent probability, though it still seems uncomfortably high!

*

For the sake of argument, though, let's consider the scenario in which we really are living in a simulation, because it offers some truly bizarre implications. For instance, we would have no way of knowing when the simulation began. Perhaps the Big Bang represented the moment when the program was initially turned on, or perhaps it all began with sim-cavemen 50,000 years ago and the programmers have been tracking the progress of humanity ever since, as some kind of experiment in evolution. It may have only begun yesterday, or an hour ago. All our memories of earlier times would be false, implanted into our minds. We could even be living in a perpetual five-minute loop of time, like a repeating history reel on display in a museum.

It's also possible that you're the only real person in the simulation. Everyone else could be a shadow being, lacking true consciousness. However, Bostrom notes that if we're weighing probabilities here, there would need to be a whole lot of single-person sim-worlds created before it became statistically more likely that you were in one of them rather than in the real world. So you shouldn't leap to the conclusion that your neighbours are artificial beings just yet.

Then there's the fun part: in a simulated world, all the rules of physics can be thrown out the window. Anything becomes possible: magic, vampires, ghosts, werewolves, superpowers, miracles. You name it. In fact, if we're in a simulation, then the programmers who created us are, to all intents and purposes, 'God'. Their powers are unlimited. They can raise us from the dead or grant us eternal life. Bostrom notes that the afterlife becomes a serious possibility.

The simulation concept has attracted an enthusiastic set of fans – perhaps unsurprisingly seduced by the lure of the anything-goes physics. The tech mogul Elon Musk has declared himself a believer, and the *New Yorker* magazine reported that two Silicon Valley billionaires are so convinced it's true, they've been funding an effort to find a way to break out into the real world.

Many of these supporters reckon the odds our world is computer generated are far higher than Bostrom himself estimated. Some regard it as a near certainty, arguing that it could explain many lingering mysteries, such as claims about supernatural phenomena, or why the universe appears to be curiously fine-tuned to support our existence.

Of course, as the simulation hypothesis has grown in popularity, it has simultaneously generated a backlash from scientists who feel that enough is enough: that, at the end of the day, it's an absurd concept and we should accept the reality of our universe.

Part of the reason they feel this way is because they believe that Bostrom and his supporters are vastly overestimating the chances that such convincing simulated worlds will ever be built. The technical challenge alone, they argue, would be daunting, and possibly insurmountable. Creating lifelike graphics in a computer game is one thing, but generating the data necessary to create an entire world, down to the quantum level, is another thing altogether. Even futuristic supercomputers might not be up to the task.

And, really, why would any advanced civilization create such a thing? The physicist Sabine Hossenfelder has made the case that anyone capable of creating artificial intelligences would surely want to put those intelligences to work solving real-world problems, not trap them in a simulated environment.

The larger complaint, though, is that the entire simulation debate seems frivolous and pointless. After all, there's nothing we can do with the information that the universe *might be* a simulation. It offers no obvious implications about how we should live our lives. Nor does it increase our knowledge in any way, because there's no way to prove or disprove the hypothesis. The whole idea, they insist, should be relegated to the category of pseudoscientific nonsense.

Bostrom himself agrees that we shouldn't alter our behaviour in any way just because we might be living in a simulation. In that respect, it's true that the hypothesis is completely irrelevant to our everyday lives.

But a line of thinking in favour of the hypothesis could be that it draws attention to one of the underlying assumptions upon which scientific knowledge is built. This is the assumption that the universe is, in the words of the astronomer William Keel, playing fair with us.

Scientists take it for granted that nature is following certain rules and isn't breaking them. One of these is that the laws of nature, such as gravity and electromagnetism, are universal and apply uniformly. But there's no way to definitively test that this is true. All we can say is that, so far, we've never observed a case where the laws of nature are broken.

Another assumed rule – and this is more directly relevant to the simulation hypothesis – is that the parts of the universe we can't see are similar to the parts that we can. When astronomers look out from the Earth, they can peer approximately forty-six billion light years in any one direction, which is the furthest light has been able to travel since the birth of the universe. Since they suspect the universe may be infinite in size, this bubble of observable space around us represents a tiny fraction of the whole universe.

Nevertheless, cosmologists routinely make conclusions about the entire universe. They can only do this because they assume the universe as a whole closely resembles the part of it we live in and can see. They assume our local cosmic neighbourhood is a representative sample of the whole thing. They refer to this assumption as the cosmological principle. But, again, there's no way to test if this is true. For all we know, once you venture past the edge of the observable universe, space might be made of mozzarella cheese.

A mozzarella universe would violate every known law of physics, but a simulated universe wouldn't break any laws. There's nothing inherently impossible about it. It's unlikely, perhaps, but not impossible. So, if we were to imagine a plausible way in which the universe wasn't playing fair with us, this might be it.

Which is to say that, if we were somehow able to zoom out and see the big-picture view of the entire cosmos, beyond the observable universe, we wouldn't find galaxy upon galaxy stretching away to infinity. Instead, we'd encounter the surface of a computer hard drive.

What if there's only one electron in the universe?

In the sixth century BC, the Greek philosopher Thales declared that everything was made from water. What exactly did he mean by this? Did he really think *everything* was made from water, or was he speaking metaphorically? Unfortunately, we're not sure, because none of his writings have survived. We only know he said this because Aristotle, writing some 250 years later, briefly mentioned that he did. The cryptic statement has nevertheless earned Thales credit for being the first person to ever utter a recognizably scientific statement, because it seems he was attempting to explain the world naturalistically rather than by appealing to mythology, which is what everyone else had done up until then. He also seemed to be suggesting that the apparent complexity of nature, all the bewildering diversity of its forms, might be constructed out of some more basic type of material.

This latter idea lies at the heart of what it means to do science. Scientists strive to understand nature by discovering the simpler patterns and structures that underlie its outward complexity, and this quest, first articulated by Thales, has yielded stunning results. Biologists have discovered that the vast assortment of living organisms on the Earth all acquire their incredible variety of forms from cellular instructions written in a genetic code that consists of only four letters: A, C, G and T. From these four letters, arranged in different ways, nature produces species as widely varied as bacteria, fungi, oak trees, polar bears and blue whales.

On an even grander scale, physicists have demonstrated that all the different substances in the universe – diamonds, granite, iron, air and so on – are made out of atoms, which, in turn, are made out of just a few types of subatomic particles, including electrons, protons and neutrons. Thales would have been impressed!

But what if this search for the hidden building blocks of nature could be taken a step further? Scientists may have identified the fundamental types of particles that all things are made out of, but what if there's actually only one thing, in a literal, quantitative sense? What if everything in existence is made from a single subatomic particle?

This is the premise of the one-electron universe hypothesis. It imagines that, instead of there being an infinite number of particles in the universe, there's only one that appears in an infinite number of places simultaneously. It achieves this trick by constantly travelling back and forth through time.

The one-electron universe hypothesis sprang from the imagination of Princeton professor John Wheeler, who was one of the most respected physicists of the twentieth century. He came up with the idea in 1940, while sitting at home one evening, puzzling over the mystery of antimatter, which, at the time, had only recently been discovered.

Antimatter is like the bizarre evil twin of matter. It presumably looks just the same as matter – no one knows for sure, because no one has ever seen a chunk of it – but it has an opposite electric charge, which means that, if the two ever come into contact, they instantly annihilate, transforming into pure energy. In fact, the mutual destruction of matter and antimatter is the most efficient way known to exist in nature of releasing energy from matter. This has apparently attracted the interest of the US military. It's long been rumoured that Department of Defense researchers have been trying to figure out how to build an antimatter bomb, the power of which would dwarf any nuclear bomb.

The existence of antimatter was first predicted in 1928 by the British physicist Paul Dirac. He had been trying to come up with an equation to describe the behaviour of electrons, but his calculations kept yielding two answers, a positive and a negative one, which struck him as curious. Most physicists probably would have ignored the negative number, assuming that it couldn't correspond to anything in real life, but Dirac believed that mathematics provided a window onto a deeper reality, even when it gave seemingly illogical answers. So, he eventually concluded that electrons must possess some kind of mirror-image subatomic doppelgänger.

It turned out he was right. The physicist Carl Anderson experimentally confirmed this in 1932 by finding evidence of exactly such an antimatter particle in a cloud-chamber experiment. The particle that Anderson detected was, in most respects, identical to an electron. It possessed the same mass and spin. However, it had a positive charge, whereas an electron has a negative charge. For this reason, Anderson called this anti-electron a 'positron'.

Anderson had confirmed that antimatter existed, but it wasn't clear what brought it into being or how it fitted into the larger picture that was emerging of the subatomic world. These were the questions John Wheeler was thinking about on that night in 1940 when he came up with his odd idea. It suddenly occurred to him that perhaps positrons were simply electrons travelling backwards in time. After all, positrons and electrons seemed to be identical in all ways except for having an opposite charge, and moving backwards in time could reverse the charge of an electron.

Wheeler imagined an electron going forwards in time, and then reversing course and coming back as a positron. This made him realize that, from the point of view of someone such as ourselves, stuck in a particular instant, it would seem as if the electron and positron were two separate particles, whereas, in fact, they were actually the same particle at different stages of its journey through time.

Wheeler then envisioned this process continuing. The electron would travel forwards until it reached the very end of time, at which point it would reverse course, travelling backwards as a positron until it bumped up against the beginning of the universe, and then it would reverse its course again. If this back-and-forth time travel continued to infinity, zig-zagging between the beginning and end of time, that single electron could conceivably become every electron in the universe. What we in the present perceive to be an enormous number of different electrons might actually be the same one repeatedly passing through our moment in time.

It occurred to Wheeler that, if this really was true, it would solve another mystery: why electrons are indistinguishable from one another. Because they are, indeed, all exactly alike. There's no possible way to tell two of them apart.

The story goes that, at this moment of epiphany, Wheeler excitedly telephoned his brilliant young graduate student, Richard Feynman, to share his revelation. 'Feynman,' he triumphantly declared, 'I know why all electrons have the same charge and the same mass.'

'Why?' replied Feynman.

'Because they are all the same electron!'

Although Wheeler had envisioned a single time-travelling electron, you can't build a universe out of just electrons. What about all the other particles, such as protons and neutrons? Are there lots of them, but only one electron?

The answer is that Wheeler's hypothesis could indeed be extended to include the whole suite of other particles. His focus on electrons alone was just an accident of history, because, in 1940, positrons were the only form of antimatter whose existence had been confirmed. It wasn't until the 1950s that researchers proved that other particles also have antimatter counterparts.

However, after sharing his weird idea with Feynman, Wheeler didn't actually try to develop his concept further. He considered it to be little more than idle speculation, and so the one-electron universe hypothesis might have faded away. That didn't happen, though, because Feynman kept it alive.

Feynman was doubtful about the hypothesis as a whole. He didn't think there could really be just a single electron in the entire universe, but he was extremely intrigued by Wheeler's idea of time-travelling electrons. He proceeded to develop this concept further, and in doing so he laid the foundations for the work that eventually made him one of the most famous physicists of the twentieth century.

What he was able to demonstrate, by the end of the 1940s, was that thinking of antimatter as time-reversed matter wasn't crazy at all. In fact, it provided a powerful way of understanding the behaviour of subatomic particles. If you envision the subatomic world in this way, then, when a particle of matter and antimatter collide, they don't actually annihilate each other in a burst of energy. Instead, the apparent collision represents the moment the particle has changed the direction of its travel through time.

This is now called the Feynman–Stueckelberg interpretation of antiparticles, and it's considered to be a perfectly valid, and widely used, way of conceptualizing antimatter. Which isn't to say that physicists believe that antimatter *really is* time-reversed matter, only that this is a useful way of modelling its behaviour. From a mathematical perspective, a positron is, in fact, the same thing as a time-reversed electron.

This lends some credibility to Wheeler's one-electron universe hypothesis, because it means that there's a nugget of valid insight at its core. Thinking of antimatter as time-reversed matter isn't some kind of crackpot idea; it's a concept that physicists take quite seriously.

Feynman received a Nobel Prize in 1965, and during his acceptance speech he told the story of Wheeler's late-night phone call to him in 1940. It was because of this speech that the one-electron universe hypothesis finally gained a wider audience.

*

If Feynman liked Wheeler's time-travelling electrons so much, however, why didn't he buy into the rest of the one-electron universe hypothesis? It was because he immediately recognized there was a big problem with the general concept of a single electron zig-zagging back and forth between the beginning and end of time. If that were true, half the universe should consist of antimatter, because this one particle would have to spend half its time going backwards (as antimatter), and half going forwards (as matter). But, as far as researchers can tell, there's almost no antimatter in the universe. Whenever antimatter is created, whether in nature or the lab, it almost immediately gets destroyed when it promptly collides with matter.

From our perspective, it's a good thing there isn't more antimatter around, as it means we don't have to worry about being instantly obliterated by stumbling upon random chunks of it. But the lack of antimatter is actually one of the great mysteries in science, because physicists believe that an equal amount of matter and antimatter should have been created during the Big Bang. If this happened, though, where did all the antimatter go? Scientists aren't sure. The current thinking is that, for various reasons, slightly more matter than antimatter must have been created during the initial moments of the Big Bang. Everything then collided in a cataclysmic event known as the Great Annihilation. After the smoke had cleared (metaphorically speaking), all the antimatter was gone, but, because of that slight initial imbalance, some matter remained, and that surviving amount represented all the matter that now exists in the universe.

This explanation, however, doesn't work for the one-electron universe, because the hypothesis suggests that the amount of matter and antimatter should be evenly distributed throughout time. During Wheeler's phone call about the hypothesis to Feynman, back in 1940, Feynman had actually pointed out the missing antimatter problem, which prompted Wheeler to suggest a possible solution: perhaps all the missing antimatter was hidden inside protons! After all, protons have a positive charge, just as positrons do. But Wheeler quickly dropped this idea as he realized that protons are about 2,000 times bigger than electrons. The size mismatch just wasn't plausible. Also, if protons were really the antimatter form of electrons, you'd expect atoms to constantly be self-destructing as their protons and electrons collided.

Fans of the hypothesis have subsequently tried to think up other solutions to the missing-antimatter puzzle. For instance, if it isn't in protons, perhaps it's hidden somewhere else, like in a distant part of the universe. Perhaps faraway stars and galaxies are actually made out of antimatter.

The idea that vast swathes of the universe might be made out of antimatter has intrigued many scientists, and astronomers have been keeping their eyes open for any evidence that this might be the case, such as cosmic fireworks where an antimatter zone is butting up against a matter-filled zone. So far, they've seen no sign of such a thing.

But there are even more exotic ways antimatter might be hiding. What if Wheeler's single particle doesn't return by the same route? In the 1950s, the physicist Maurice Goldhaber suggested that an antimatter universe might have formed alongside our matter-based universe. After all, if subatomic particles come in matter/antimatter pairs, why shouldn't this also be true of the universe as a whole? If such an anti-universe exists, perhaps the particle returns to the beginning of time via this route.

Or perhaps time loops. Instead of having to go back the way it came, perhaps, when an electron reaches the end of time, it instantly finds itself returned to the beginning.

Critics of Wheeler's hypothesis complain that speculations of this kind roam pretty far from any kind of verifiable evidence, into the realm of pure whimsy. They also ask, 'So what?' Even if the hypothesis were true, what new knowledge would we gain? What new possibilities for research or understanding would it open? Seemingly none, because, whether the universe is filled with an infinite number of particles or one particle in an infinite number of places simultaneously, the two amount to the same thing. The physics is the same.

Though, a possible gain would be to fulfil that ancient scientific dream of Thales by identifying the ultimate building block of nature. And don't discount the wow factor. It would certainly give new meaning to the idea of being one with the universe.

What if we're living inside a black hole?

If you ever happen to fall into a black hole, you'll be spaghettified. Yes, this is the actual scientific term for what will happen. The crushing gravity of the black hole will simultaneously stretch and compress you, moulding you into a stream of subatomic particles resembling long, thin noodles. These particles, which once were you, will then plummet down to the centre of the hole, where they'll be further compacted to a density so extreme that mathematics can't even quantify it.

The one silver lining in all of this is that, if you happened to be conscious when you fell into the black hole, the spaghettification would occur so quickly that you'd scarcely feel a thing. All in all, not a bad way to go.

Given that black holes are such hostile environments, they would seem to be unlikely places for life to exist. After all, what could possibly survive in them? Well, there's a popular theory that contradicts this logic. It argues that, improbable as it might seem, not only can life exist in such a place, but we're proof of it, because the universe we're living in is a giant black hole.

A black hole is defined as an object whose gravity is so extreme that nothing can escape from it, not even light. The suspicion that such objects might exist dates back several centuries, though for most of that time scientists were unwilling to accept that a phenomenon so bizarre could possibly be real. The first person to predict their existence was the English clergyman John Michell. He submitted a paper to the Royal Society in 1783 in which he speculated that, if a star were 500 times larger than our own sun, the strength of its gravitational field would prevent light from escaping. Despite its vast size, the star would disappear from sight, becoming a black hole (although Michell didn't use that term).

The scientific community dismissed this hypothesis as wild conjecture. It didn't seem plausible that any star could be that large. And, anyway, the prevailing belief was that light wasn't affected by gravity. So, the odd idea of black holes was filed away and forgotten.

It wasn't dusted off and revived until 1915, when Albert Einstein published his general theory of relativity. This convinced scientists that gravity wasn't a force, but instead represented the curvature of space and time. In which case, light *would be* affected by gravity, because it would follow the curve of space–time as it travelled.

General relativity also implied that an object of sufficient mass and density might distort space–time so severely that it would form a well from which nothing, not even light, could climb out. At the centre of this well, a singularity would form – a point at which the strength of the gravitational field became infinite.

Even though this is what the theory implied, most scientists, including Einstein himself, continued to regard black holes as a crazy idea. The problem was that the laws of physics would break down in a singularity, and physicists instinctively shied away from accepting that this could happen. The idea that matter could be infinitely compressed also seemed wrong, because how can something that's finite acquire an infinite value? Scientists assumed that, at some point, the subatomic particles making up matter would find a way to resist further compression.

It wasn't until the 1960s that the scientific community finally came around to accepting the existence of black holes. In fact, it was only in this decade that the term was coined; science journalist Ann Ewing gets credit for first using it in a 1964 article. The complex mathematics of how space would curve around black holes had come to be better understood, which made physicists feel more comfortable with the whole idea of them. Also, the new technologies of radio and X-ray astronomy were revealing that the universe contained some very strange, high-energy and extremely dense objects, such as quasars and pulsars. By comparison, black holes no longer seemed as improbable.

Once black holes had been accepted as a plausible phenomenon, it didn't take long for people to start wondering if we might be living inside one.

The black-hole universe hypothesis wasn't the brainchild of a single theorist. There was no one figure who emerged as its champion. Instead, it was a concept that began circulating like a meme within the scientific community during the late 1960s and early 1970s, and then caught the popular imagination.

The physicist Roger Penrose may have been the first to speculate about the idea in print. He mentioned the possibility in a 1967 essay he submitted for Cambridge University's Adams Prize. The hypothesis reached a wider audience five years later when the physicists Raj Pathria and Irving Good each authored a brief article about it, which appeared, respectively, in the journals *Nature* and *Physics Today*. By the 1980s, the hypothesis was well established as a trendy, if unorthodox, idea, and discussions of it regularly appeared in books and articles.

The hypothesis itself occurred independently to a number of people because, within the context of astrophysics, it's actually a somewhat obvious idea to arrive at. There are only two places in nature associated with singularities: the centre of a black hole and the start of the Big Bang that created our universe. So, it's logical to wonder if the two might be related.

Once you start comparing the features of our universe with a black hole, other similarities suggest themselves. There's the fact that a black hole has an event horizon. This is a point of no return, an invisible line around the black hole that, once crossed, forbids exit for anything, even light. Whatever passes over that line is entirely within the clutches of the gravity of the black hole and, from the point of view of an external observer, it vanishes completely, effectively ceasing to exist as part of the universe. The event horizon acts as a barrier. Nothing that is within its circumference can ever travel outside of it.

Similarly, we're trapped within a cosmological event horizon defined by the limits of how far we can see out into the universe (about forty-six billion light years in each direction). We can never travel beyond that horizon. We're trapped inside it just as completely as objects within the event horizon of a black hole are trapped.

The reason we can't travel beyond our cosmological horizon is because the universe is expanding, and this causes the horizon to move away from us faster than we can travel to catch up with it. It's actually receding faster than the speed of light, and, although it's not bound by light's speed limit, we are. This means that the laws of physics forbid us from ever travelling beyond that horizon.

This all may sound contradictory. Why can't we travel faster than the speed of light, when the cosmological horizon can? It's because space is expanding, and it's doing so everywhere, which means that the expansion is cumulative. The more distance lies between two objects, the more units of space there are that are simultaneously expanding, without limit. Add together enough points of space and the expansion rate will eventually surpass the speed of light. No one ever said astrophysics was simple or intuitive!

Then there's the issue of the Schwarzschild radius. In 1915, soon after Einstein published his theory of general relativity, the German physicist Karl Schwarzschild used its equations to compute what the strength of the gravitational field would be surrounding any uniform ball of matter. This was an impressive feat, not only because Einstein's general relativity equations are famously difficult to solve, but also because Schwarzschild was in the German army, dodging bullets on the Russian front, when he did this. Plus, he had contracted an incurable skin disease which was rapidly killing him. He mailed his calculations off to Einstein, and then he died.

Schwarzschild's analysis came to a seemingly bizarre conclusion. It suggested that any object would become a black hole if it was squeezed small enough, because the pull of gravity at its surface would increase as its mass grew denser. The strength of an object's gravity is inversely related to your distance from it – specifically, to the distance between you and the centre of its mass. So, if you reduce the distance to its centre by compressing the entire mass into a smaller diameter, its gravity will grow correspondingly.

Let's look at an example. If mad scientists were able to take the entire mass of the Earth and compress it down to slightly smaller than a golf ball, the pull of gravity on its surface would become inescapable. It would turn into a black hole. Similarly, if these scientists could shrink you down to a tiny speck, smaller than the nucleus of an atom, you'd also become a black hole. The size at which an object becomes a black hole is now known as its Schwarzschild radius, and this radius can be calculated for any object.

Scientists initially dismissed Schwarzschild's finding as a theoretical oddity, because they were reluctant to believe that black holes were a genuine phenomenon. But, once they had come around to accepting their reality, it occurred to some to ask what the Schwarzschild radius of the observable universe might be. Which is to say, how small would you need to shrink the observable universe in order to transform it into a black hole?

The mass of the observable universe can be estimated by observation, and we know its size. (We can see forty-six billion light years in any direction, so it's ninety-two billion light years across.) When these numbers were run through the equations, the disturbing result emerged that the observable universe lay within its Schwarzschild radius. It was already at its black-hole size!

This conclusion may sound impossible, because surely an object must be enormously dense to have the gravity of a black hole, and yet, looking at the universe around us, there's plenty of empty space. But this highlights another peculiarity of Schwarzschild's calculations. His analysis revealed that the more massive an object is, the less dense it needs to be to become a black hole. For instance, if those mad scientists were able to compress the entire Milky Way galaxy within its Schwarzschild radius, its density would be less than that of water in the ocean. And the mass of the entire observable universe, confined within its Schwarzschild radius, would not be very dense at all. In fact, it would be exactly as dense as we observe it to be.

So, the argument goes that, if you consider all these factors together – the singularity associated with our universe through the Big Bang, its cosmological event horizon and the fact that our observable universe lies within its Schwarzschild radius – you're led to the seemingly inescapable conclusion that we must be living inside a black hole.

Of course, most astrophysicists aren't about to concede this. For a start, they note that the singularity of the Big Bang isn't comparable to the singularity of a black hole, because it's in the wrong place. If you were to fall into a black hole, the singularity would lie unavoidably ahead of you, in your future, but in our universe the singularity lies in our past, when the Big Bang occurred. That's a big difference. Our universe seems to have emerged from a singularity, but it's not heading towards one.

Also, the event horizon of a black hole is a fixed boundary in space, whereas the cosmological horizon is relative to one's position. A civilization twenty billion light years away sees a distinctly different cosmological horizon than we do.

And, as for the observable universe being within its Schwarzschild radius, it's true that it is. This doesn't necessarily mean we're in a black hole, though. What it means is that the rate of expansion of our universe has closely matched its escape velocity, and we should be thankful for this. If the universe had expanded slower, it might have collapsed back in on itself, and, if it had expanded faster, it would have been impossible for structures such as galaxies and solar systems to form. Instead, it's been expanding at just the right rate to allow us to come into existence.

The theoretical physicist Sean Carroll has pointed out that, if one is really keen on the idea that we're stuck in some kind of gigantic cosmic hole, a better argument might be made for the idea that we're living in a white hole, which is the opposite of a black hole. More specifically, it's a time-reversed black hole out of which matter spews unstoppably rather than falling inwards. But white holes come with their own set of problems. They're theoretically possible – because the laws of physics work equally well whether one goes forwards or backwards in time – but only in the same way that it would be theoretically possible for a broken egg to spontaneously reform into a whole egg. Physics allows it, but the odds of ever seeing such a thing happen seem close to zero.

So, perhaps we're not living in a black hole. The fact alone that our universe isn't collapsing inwards towards a singularity surely proves this. But, in astrophysics, things are never quite that simple. Fans of the black-hole universe hypothesis insist there are ways around all the contrary arguments.

One way it could still work is if a singularity does lie inescapably in our future. If, billions of years from now, the expansion of the universe were to halt and then reverse, leading to a 'big crunch', this, they say, might satisfy the definition of a black hole.

Theoretical physicist Nikodem Poplawski has proposed another way. He argues that, inside a black hole, matter might actually expand outwards, rather than contracting inwards. This could be the case if matter reaches a point at which it can be crushed no further, and it then bounces back out explosively, like a coiled-up spring. The event horizon would continue to separate the contents of the black hole from the wider universe, and the expansion would then occur into an entirely new dimension of time and space, exactly resembling the sudden expansion of time and space during the Big Bang. Poplawski goes so far as to propose that all black holes form new universes, and that this is the way our own universe came into existence, as a black hole that formed in some larger universe.

Of course, this raises the question of how the parent universe might have formed, which would seem to be a mystery. Unless it too was originally a black hole. Perhaps the cosmos consists of a series of black-hole universes, nested within each other like Russian dolls, extending infinitely all the way up and down. A mind-boggling thought, perhaps, but arguably not inherently more far-fetched than any other theory about the origin of our universe.

Weird became true: dark matter

When you look up at the night sky, you see thousands of stars twinkling above you, but mostly you just see darkness. You might imagine this blackness is empty space, but not so. Astronomers now believe it's full of invisible 'dark matter'. Particles of this stuff may be drifting through you right now, but you'd never be aware of it because dark matter is fundamentally different from ordinary matter. The two barely interact at all. There's also way more of it than there is matter of the visible kind – over five times as much.

Given the oddness of this concept, it shouldn't come as a surprise that, for a long time, astronomers themselves were reluctant to believe in dark matter. In fact, there was a gap of almost half a century between when its existence was first proposed and when mainstream science accepted it as real.

The man credited with discovering dark matter was Fritz Zwicky, a Swiss astrophysicist who moved to the United States in 1925 after accepting a position at the California Institute of Technology. He spent part of his time there doing sky surveys; this is routine astronomical work that involves cataloguing objects throughout the cosmos. In the course of this activity, his attention was drawn to the Coma Cluster, which is an enormous group of galaxies located about 320 million light years away from Earth.

Zwicky noticed that the galaxies in this cluster were moving extremely fast – so fast that, by his reckoning, they all should have scattered far and wide long ago. Instead, they remained gravitationally bound to each other as a cluster, and this puzzled him because he knew it would require a lot of gravity to counteract their speed. He did the calculations to figure out exactly how much, and that's when things got weird. He estimated it would take as much as fifty times more gravity than all the visible galaxies in the Coma Cluster together could produce.

Most astronomers probably would have assumed they had made a mistake in their calculations, but Zwicky was confident in his work, and he also wasn't shy about leaping to grand theoretical conclusions. He decided there had to be massive amounts of hidden extra matter that was keeping the entire cluster gravitationally bound together. In a 1933 German-language article, he described this as 'dunkle Materie' or 'dark matter'.

Zwicky's theory fell on deaf ears. This was partly because he had published in a German journal, which meant it fell under the radar of much of the English-speaking astronomical community. But, more significantly, his idea was outrageous. It wasn't controversial to claim there was stuff in the cosmos that optical telescopes couldn't see, such as planets or burned-out stars, but Zwicky was claiming there was far more of this dark matter than there was visible matter. Huge amounts more. Naturally, astronomers wanted more proof before accepting such a radical concept.

Zwicky also found it difficult to find allies willing to take his strange idea seriously for a more prosaic reason – because his colleagues didn't like him much. He had a reputation for being abrasive, cranky and highly opinionated, often referring to those who dared to disagree with him as 'spherical bastards' (his term for a person who was a bastard from whatever angle you looked at them). Behaviour of this kind didn't endear him to his colleagues. As a result, dark matter languished in obscurity. It wasn't until the 1970s that the concept finally emerged into the light, thanks to the work of Vera Rubin.

Like Zwicky, Rubin experienced social challenges in having her work accepted, but this, more frustratingly, was due to her gender. When she began her career, she was one of the few female astronomers and had to struggle to be taken seriously in the male-dominated profession. Where Zwicky raged at those who disagreed with him, Rubin patiently collected more and more data until it eventually became impossible to ignore what she was saying.

Rubin studied individual galaxies, rather than entire clusters, but, like Zwicky, she noticed something odd about their motion. The outer arms of the galaxies she examined were moving too fast. At the speed they were going, the arms should have spun off entirely, unless there was some extra gravitational mass keeping them on. In 1970, she published an article with the astronomer Kent Ford in which they noted that, in order to account for the speed of rotation of the arms of the Andromeda Galaxy, it needed to contain almost ten times as much mass as was visible.

In the following years, Rubin and Ford continued to produce similar observational data for more galaxies, and other researchers confirmed their findings. As a result, by the end of the 1970s, scientific opinion had swung decisively in favour of the existence of dark matter. The consensus was that there was simply no other explanation for all this data.

But what exactly was dark matter? Zwicky had assumed it was just non-visible regular matter, but, if that was the case, there should have been various ways to detect its presence other than by the effect of its gravity. Astronomers tried all these techniques, but they all came up blank. So, gradually, the belief grew that dark matter had to be made of something more peculiar.

A whole laundry list of candidates have been considered and rejected, leading to the current most popular hypothesis, which is that dark matter must be some kind of as-yet-unknown type of subatomic particle that doesn't interact with regular matter except through gravity. Presumably this stuff is all around, but it concentrates in vast halos around galaxies, providing the scaffolding that allows them to form.

However, all attempts to directly detect dark-matter particles have, so far, failed, leading some sceptics to question whether the stuff actually exists. They've suggested that the effects being attributed to dark matter might actually be caused by gravity somehow operating in a different way on galaxy-sized scales, as opposed to smaller, solar-system-sized scales. This alternative theory, first put forward by the Israeli physicist Mordechai Milgrom in 1983, is called Modified Newtonian Dynamics, or MOND, and its models work surprisingly well to explain the motion of individual galaxies, though far less well to account for the motion of entire clusters of them.

To most scientists, however, the idea of revising the law of gravity borders on heresy. For which reason, MOND has attracted only a handful of supporters. But, as long as dark-matter advocates can't completely close the deal by detecting particles of the stuff, hope remains alive for MOND. And, the longer dark matter eludes direct detection, the more doubts about it will come to the fore. Which means that although Zwicky's weird idea has now been embraced by the scientific mainstream, there's a small but real chance that one day its fortunes could change again, which would make it a weird theory that became true, but then became no longer true after all.

What if we live forever?

Have you ever had a close brush with death? Perhaps you were crossing a street, one foot off the curb, when a car sped by and missed you by inches. Perhaps an object fell from a tall building and almost hit you, but instead crashed to the ground a few feet away. Or perhaps you were deathly ill, but made a miraculous recovery. The varieties of potential near-misses are endless.

Here's a disturbing thought: perhaps you did die. Or rather, you died in one version of reality, while, in another (your here and now), you remained alive. This possibility is suggested by one of the strangest theories in physics: the many-worlds theory of Hugh Everett. It imagines that everything that can happen does happen, because the universe is constantly splitting into parallel realities in which every possibility is realized. So, all near-death experiences should produce outcomes both in which you survive and in which you don't. These scenarios will exist simultaneously.

Everett's theory may sound more like science fiction than actual science. It does, regardless, offer an elegant solution to several perplexing problems in both physics and cosmology. This has earned it the endorsement of a number of leading scientists. It might have won over even more supporters if it weren't for the seriously bizarre implication that, if everything that can happen does happen, then, surely, in at least a handful of the many parallel worlds that might exist, we're all going to find a way to keep cheating death and live forever.

Everett's theory emerged from the discipline of quantum mechanics, which was formed in the early twentieth century as researchers began to gain an understanding of the mysterious subatomic world. What physicists realized as they explored this realm was that, to their utter astonishment, the rules governing the behaviour of subatomic objects were very different from the rules governing objects in the everyday world around us. In particular, it became apparent that subatomic particles such as electrons could be in more than one place at the same time. In fact, they could be in many places simultaneously.

This phenomenon is called superposition, and, if you're not familiar with quantum mechanics, it may sound strange, but physicists have no doubt about it. They believe it to be absolutely real. Nowadays, it's even being employed in real-life applications, such as quantum computers, which use the principle to perform multiple calculations simultaneously, giving them unprecedented speed. Superposition is, however, exactly as weird as it sounds.

Physicists were led to accept its reality as they struggled to predict and understand the movement of subatomic particles. In classical mechanics, if you fire a bullet out of a gun, you can predict exactly where it will hit. Its trajectory follows very logical rules. In the subatomic world, however, there's no such certainty. Researchers realized that, when they fired a photon (a particle of light) or electron out of a gun, there was no way to predict exactly what trajectory it would take and where it would hit. It simply couldn't be done.

What they *could* do was map out the probability of where these particles would hit, but this added a new layer of confusion because it turned out that this map didn't form anything like the pattern they expected. Instead, it indicated that subatomic particles were moving in very illogical ways. For example, when they fired a beam of photons at a metal plate perforated by two slits, at a rate of one photon per second, what they expected was that a strike pattern of two straight lines would form on the wall behind the slits. This is what would happen if bullets were fired at a double-slitted metal plate. But, instead, the photons created a wave-interference pattern, like two ripples colliding in a pond would make.

By the rules of classical mechanics, this should have been impossible. Researchers couldn't make head nor tail of what was happening, until they considered the possibility that each individual photon was moving through both slits simultaneously. As counter-intuitive as it seemed, the wave-interference pattern had to be the result of each photon interfering with itself.

To visualize this, imagine a photon simultaneously following every possible trajectory it can take. All these paths fan out like a wave that hits both slits. As the wave passes through the slits, it produces an interference pattern on the other side. All its potential trajectories coexist in superposition, interacting with each other like ghostly invisible lines of force.

This concept was put into mathematical form in 1925 by the German physicist Erwin Schrödinger. He devised an equation that predicted the behaviour of quantum systems by plotting this 'wave function' of subatomic particles, producing a map of every possible trajectory a particle might take. Physicists continue to use Schrödinger's equation to model the behaviour of quantum-mechanical systems with very high accuracy.

But there was a catch. The concept of superposition worked brilliantly to explain the seemingly bizarre way that subatomic particles were moving, but it didn't explain why, when a particle eventually hit the wall behind the double-slitted plate, it actually only hit it in one place. This defied the claim that the particle was following multiple trajectories simultaneously. How did the particle go from being in superposition to being in just one position?

This was the question that perplexed physicists as they came to accept the reality of superposition. It was the enigma of how particles transformed, from probabilistic objects existing in multiple locations simultaneously, into definite objects fixed in one position in space. This came to be known as the measurement problem, because the transformation seemed to occur at the moment of measurement, or detection.

In the 1920s, the physicists Niels Bohr and Werner Heisenberg came up with a solution to the measurement problem. They proposed that it was the act of observation which, by a method unknown, caused the infinite number of possible trajectories described by Schrödinger's wave function to collapse down into just one trajectory. By merely looking at a particle in a state of superposition, they argued, an observer caused it to select a single position. Because Bohr was Danish and lived in Copenhagen, this came to be known as the Copenhagen interpretation of quantum mechanics.

The interpretation had some odd philosophical implications. It suggested that reality didn't exist unless it was observed – that we, as observers, are constantly creating our own reality somehow from the waves of probability surrounding us. Despite this weirdness, the majority of the scientific community rapidly accepted the Copenhagen interpretation as the solution to the measurement problem. This might have been because Bohr commanded enormous respect. No one dared contradict him. Plus, there didn't seem to be any other compelling solution available.

Not all scientists, however, were happy with Bohr and Heisenberg's solution. It seemed ludicrous to some that the act of observation might shape physical reality. Einstein himself was reported to have complained that he couldn't believe a mouse could bring about drastic changes in the universe simply by looking at it.

Hugh Everett, a graduate student in physics at Princeton during the early 1950s, was among these sceptics. The entire premise of the Copenhagen interpretation seemed illogical to him. He couldn't understand how a subatomic particle would even know it was being observed. Then, one night, as he was sharing a glass of sherry with some friends, a different solution to the measurement problem popped into his head. It occurred to him that perhaps the strange phenomenon of superposition never actually disappeared upon being observed. Perhaps the wave function never collapsed. Perhaps all the possible trajectories of a particle existed *and continued to exist* simultaneously in parallel realities. It just seemed to us observers that the wave function collapsed because we were unable to perceive more than one of these realities at a time.

Everett quickly became enamoured of his idea and decided to make it the subject of his doctoral dissertation, which he completed in 1957. His argument, as he developed it, was that the Schrödinger equation wasn't just a mathematical equation. It was a literal description of reality. Every possible trajectory that the wave function described was equally real, in a state of superposition – as were we, the observers. Therefore, when a researcher measured a particle, multiple copies of his own self were viewing every possible trajectory of that particle, each of his copies thinking it was seeing the only trajectory.

Carrying this argument to its logical conclusion, Everett posited that the fundamental nature of the universe was very different from what we perceived it to be. Our senses deceived us into believing that there was only one version of reality, but the truth was that there were many versions – a vast plurality of possible worlds – existing simultaneously in superposition.

What this implied was that anything that physically could happen must happen, because every possible trajectory of every wave function in the universe was unfolding simultaneously. This didn't allow for the existence of supernatural phenomena, such as magic or extrasensory powers, because these aren't physically possible. But it did suggest that there were versions of the universe in which every physically possible scenario played out. Somewhere out there, in the great quantum blur through which our consciousness was navigating, there had to be versions of reality in which the Earth never formed, life never began, the dinosaurs never went extinct and every single one of us won the mega-lottery. Every possible chain of events, no matter how improbable, had to exist.

Initially, the scientific community ignored Everett's dissertation. For thirteen years, it languished in obscurity. In response to this silent rejection, Everett abandoned academia and took a job with the Pentagon, analysing the strategy of nuclear weapons. He never published again on the topic of quantum mechanics. But Everett's theory did eventually catch the attention of cosmologist Bryce DeWitt, who became its first fan. Thanks to his promotional efforts, which included republishing Everett's dissertation in book form and coining the name 'many-worlds theory', it reached a wider audience.

And, once they became aware of it, physicists didn't dismiss Everett's idea out of hand. Predictably, many of them bristled at the idea that we're all constantly splitting into parallel copies. After all, the world around us seems reassuringly solid and singular. But the theory did manage to find a handful of converts, who noted that, as a solution to the measurement problem, it worked, and it did so without having to invest magical powers in the act of observation, as the Copenhagen interpretation did.

Some cosmologists were also intrigued by it. They noted that it could explain the 'fine-tuning' problem that had recently come to their attention. This problem was that, in order for life to be possible, there are hundreds of aspects of the design of the universe that had to be fine-tuned just right during the Big Bang. For instance, it was necessary for protons and neutrons to have almost the same mass, for the relative strengths of electromagnetism and gravity to be exactly as we observe them to be, and for the universe to be expanding at precisely the speed that it is. If any of these values had been different, even just slightly, life would never have been possible, and there are many more constants and ratios of this kind that had to be perfectly calibrated. All of these values, however, seem somewhat arbitrary. It's easy to imagine they could have been different. So why did they all come in at exactly the right numbers needed for life to emerge?

If there's only one universe, the odds of life existing seem beyond incredible. It would be like rolling double sixes a million times in a row. But if there are many parallel universes, in which all physical possibilities occur, then some of them are bound to be appropriately fine-tuned to allow the emergence of life.

A technical issue, however, still made many physicists reluctant to accept Everett's theory. They couldn't understand why all the trajectories of a particle would become independent of each other. Why would we ever perceive ourselves to be locked in one reality? Why didn't all the many worlds remain blurred together, as one?

In the 1980s, the physicist Dieter Zeh, of the University of Heidelberg, developed the theory of decoherence, which provided a possible answer. It hypothesized that the wave function of a particle interacts with the wave functions of surrounding particles, and, as it does so, it tends to 'decohere'. Particle trajectories entangle with each other, tying up, as if in knots, and this causes the various parallel worlds to become independent of each other.

Decoherence was a highly sophisticated mathematical theory, and, when coupled with Everett's many-worlds theory, the two provided a compellingly complete model of the underlying subatomic reality we inhabit. As a result, support began to shift away from the Copenhagen interpretation and towards the many-worlds theory, and this trend has continued to the present. One reason the many-worlds theory continues to be resisted by most scientists, though, is because it leads to such bizarre implications. In particular, there's that immortality feature.

The first acknowledgement in print of the death-defying implication of the many-worlds theory appeared in 1971. The physicist Mendel Sachs wrote to the journal *Physics Today* noting that, if Everett's theory was true, it could cheer up a passenger on an aeroplane about to crash, because he could reflect that in some other branch of the universe the plane was certain to land, safe and sound.

By the 1980s, it had sunk in among scientists that it wasn't just a temporary escape from death that the many-worlds theory promised, but full-blown immortality. After all, if whatever can happen, does happen, then every time we face the possibility of death, at least one version of our self must find a way to carry on, because there's always some combination of events that would save us. Our chances of survival will grow increasingly improbable as time goes on, but improbable is not the same as impossible. In an obscure branch of the quantum universe, we'll live forever.

In 1998, the physicist Max Tegmark, one of Everett's most vocal champions, pointed out an interesting aspect of this immortality feature: it provides us with a way to test the many-worlds theory so we can know for sure whether it's true or false.

Tegmark's idea was to design a gun that would randomly either fire a bullet or merely make an audible click each time someone pulled the trigger. The chance of either outcome was fifty–fifty. The experimenter would then place his head in front of the barrel and instruct an assistant to fire. Tegmark figured that, if the gun clicked instead of firing ten times in a row, this would be ample proof of the validity of the many-worlds theory.

This 'quantum suicide' experiment is well within the scope of available technology. The problem, of course, is that no one in their right mind would want to be the guinea pig, and no one would actually believe the experimenter even if they survived.

Philosophers have also been intrigued by the concept of quantum immortality. There's now a small genre of philosophical literature devoted to exploring its practical and ethical implications. One of the ongoing debates is about whether it would be rational for a believer in the many-worlds theory to play Russian roulette, assuming a large monetary award was involved. After all, if the theory is true, at least one copy of the believer is guaranteed to win the bet every time. The other copies will be dead and therefore beyond caring.

The philosophical verdict on this question is mixed. It's true, some acknowledge, that you might make a few of your future selves wealthier, but most point out that there are many negatives to consider. Even if you don't care about killing off some of your future selves, what about the friends and family of those selves that will suffer the loss? Also, there's the simultaneous guarantee that some of your future selves will survive in a diminished capacity, either with brain damage or half their face blown off. It seems a high price to pay for the knowledge that, in some parallel world, you're slightly richer.

Everett himself never wrote about quantum immortality, but he was apparently aware of its possibility. One of his work colleagues, Keith Lynch, reported that he once discussed the idea with Everett, who declared himself to be a firm believer in the concept. This may have been what he believed, but in our world Everett died of a heart attack in 1982, at the relatively young age of fifty-one – a victim of years of poor diet and lack of exercise.

On a darker note, his daughter Liz, who struggled with addiction for much of her life, committed suicide in 1996 at the age of thirty-nine. She left behind a note in which she asked her family to throw out her ashes in the trash, as her father, a lifelong atheist, had similarly requested for himself. That way, she wrote, 'I'll end up in the correct parallel universe to meet up w/ Daddy.'

Critics view quantum immortality as a self-evidently preposterous notion that casts doubt on the entire many-worlds theory. It's hard to argue against this. We don't ever encounter immortals wandering around. Everyone, it seems, dies sooner or later. And yet, the many-worlds theory does explain the mystery of superposition very well. What better explanation is there? This has led some of its supporters to seek a way to remove quantum immortality from the mix. With it gone, the thinking goes, Everett's theory would become that much more credible.

Tegmark, for example, has argued that immortality would only be guaranteed if life-and-death events could always be reduced to a set of binary possibilities, such as a car speeding by that either hits you or doesn't. It's a situation that has only two outcomes. The ageing process, however, isn't binary; it involves millions of cumulative events building upon each other, and this might eventually overwhelm the odds, making death inevitable even in a many-worlds universe.

If Tegmark's take on the many-worlds theory is correct, some of our parallel selves might live a long time, but they'll all perish sooner or later, experiencing a variety of different deaths. But what if he's wrong? What if we all really are going to live forever?

That's the interesting thing about the many-worlds prediction of immortality. You don't have to conduct a quantum-suicide experiment to determine if it's true. In fact, you don't need to do anything at all. If Everett was right, in due course you're guaranteed to find out.

CHAPTER TWO

A Pale Blue Peculiar Dot

We've just looked at the universe in its entirety. Imagine it before us, stretching out on either side to infinity. Astronomers believe that, from a vast distance (if it were somehow possible to see it from the outside), it might appear smooth and featureless. But now let's move in closer. As we approach, hints of structure emerge. We can see the dim outline of millions of galaxies. They're not evenly spaced out. Instead, they're arranged into vast shapes, like filaments and walls that span up to a billion light years in length, criss-crossing space like a web. And within these structures there are countless smaller knot-like groups of galaxies.

We'll direct our attention towards one of these groups, the supercluster Laniakea. It's over 500 million light years across. So, we're still seeing the universe on an unimaginably vast scale. But, zooming in faster, one galaxy near the outer edge of Laniakea is our target: the spiral-armed Milky Way. It's a mere 100,000 light years across, although it may contain as many as 400 billion stars. No one knows the exact number because it's not possible to count them all. We can only make a guess based on the estimated mass of the galaxy.

As our descent continues, we head towards one of the smaller arms of the Milky Way, the so-called Orion Arm, and here, about two thirds of the way from the centre of the galaxy, we finally arrive at our destination: a medium-sized yellow dwarf star that has eight planets in orbit. The third planet is particularly striking. It's like a brilliant blue, green and white marble suspended in the vacuum of space. This, of course, is our home: the solar system and Earth.

This region, which is studied by astronomers and planetary scientists, will be our focus in this section. You might think that because we're in a more familiar setting the theories about it will be somewhat more restrained, but, as we'll see, the solar system is no less full of mysteries and controversies than the cosmos itself.

What if the Earth is at the centre of the universe?

Where is the Earth located in the universe? If you had asked scholars this question 500 years ago, they would have had a ready answer. It's right at the centre! The sun, moon, planets and stars all revolve around it. This had been the accepted belief for over 2,000 years, since the ancient Greeks, Egyptians and Babylonians had first begun to study astronomy.

But ask astronomers the same question today and the answer is no longer as simple. They'll definitely tell you we're not at the centre; they're sure of that. In fact, the idea has come to seem hopelessly naive – a product of a time when people simply didn't know as much about the cosmos and so assumed that humans occupied a privileged place in it.

If you press for a more specific answer, astronomers might give you our location in relation to other things around us. The Earth, they'll say, is the third planet from the sun in the solar system, which is part of the Milky Way Galaxy, which in turn neighbours the Andromeda Galaxy. Both of these are part of the Laniakea supercluster, which is home to some 100,000 galaxies.

And what if you keep pressing, asking for the Earth's absolute location in the universe? Are we near its top or bottom? It's southern edge, perhaps? This is where things get more complicated. Astronomers will explain that there actually is no such thing as an absolute location, because the universe doesn't have an overall shape or structure against which we could orient ourselves. On a cosmic scale, there are no large-scale features, such as a boundary or a centre, from which we could derive our position.

But what if there actually are? What if the universe does have an overall structure that can provide some orientation, and, when we locate ourselves with respect to this, the Earth turns out to be right at the centre? This was the curious argument made by the South African cosmologist George Ellis in an article published in the journal *General Relativity and Gravitation* in 1978.

Ellis wasn't some kind of crackpot. If he had been, critics could have filed his idea away with the rants of flat-Earth theorists and other members of the lunatic fringe. To the contrary, he had established his credentials as one of the foremost cosmologists in the world by co-authoring *The Large Scale Structure of Space–Time* with Stephen Hawking. This book, which came out in 1973, is considered to be one of the classic texts of cosmology.

In line with not being an oddball, Ellis didn't claim that the Earth was at the centre of the solar system. That would have been really crazy. Instead, he was concerned about the Earth's position in the universe as a whole – the big picture of our place in the cosmos. Not that this made his argument any more palatable to cosmologists.

Ellis's argument was relatively straightforward. He noted that observational data had led astronomers to conclude that the universe has no overall structure and no centre. But this same data, he maintained, could potentially be used to produce a model of the cosmos that does have a centre, with the Earth located right there. So, if the same data could yield two different but equally valid models, then the choice of which one to adopt had to be seen as a matter of philosophical preference rather than of scientific necessity.

So, what were these observations that had led astronomers to decide the universe was centreless? More than anything else, it was Edwin Hubble's discovery that the universe is expanding. In the

1920s, a powerful new telescope installed at California's Mount Wilson Observatory had allowed him to discern that the blurry nebulae in the night sky, whose nature had long puzzled astronomers, were actually distant galaxies, and, upon closer inspection, he realized that almost all these galaxies, in every direction, were rushing away from us at enormous speeds.

A simplistic interpretation of this observation would have been that the Milky Way was at the centre of the universe and that every other galaxy was being flung away from it by some powerful force. But then, why would this force be acting on everything except the Milky Way? That didn't make sense. Instead, astronomers decided that the recession of the galaxies had to mean that everything in the universe was expanding away from everything else. And it was this conclusion that eliminated the centre from the universe.

The analogy of a rubber balloon is often used to describe the concept. It asks us to imagine small paper dots glued onto a balloon, the surface of which stretches as it inflates, causing all the dots to move away from each other. The bigger the balloon grows, the more space separates the dots. If you were a microscopic person standing on one of the dots, you'd look out and see the other dots all receding into the distance, which might lead you to think that your dot occupied a central position on the surface of the balloon. But that interpretation would be wrong, because the balloon's surface has no centre. (Pretend it doesn't have an opening for air.) The view from every dot is the same. No matter which dot you stand on, you'll always see other dots receding into the distance.

In the same way, astronomers explain, the expanding universe has no centre. When we look out at the cosmos, we see galaxies receding from us on all sides, but, wherever we might happen to be in the universe, we'll always see the same phenomenon, because everything is expanding away from everything else.

The balloon analogy is potentially misleading, because a balloon does have an inside and an outside, which the universe doesn't. But, if you can somehow imagine a balloon that lacks these spatial qualities, but which is expanding at all points, then this is the model of the universe that came to be widely accepted by scientists in the 1930s, and continues to be accepted today.

What Ellis proposed as an alternative to the expanding-balloon universe was a universe shaped like an egg (though he didn't call it that). He named it the 'static spherically symmetric' universe. But it was, in essence, an egg universe, standing upright.

Each pointy end of the egg represented a centre – or rather, a centre and an anti-centre. At the top of the egg, there was a cool, relatively empty region of space, while at the bottom there was a naked singularity, a region of ultra-dense high-gravity matter, like that found at the centre of a black hole, but with no event horizon. In between the top and bottom, rings of galaxies floated, more spaced out towards the top, but increasingly dense as they neared the singularity at the bottom.

This is a gross simplification of his model, which was far more theoretically sophisticated. To put it in a way closer to his description, he envisioned a universe with a very high concentration of galaxies at the centre and a sparser concentration around the edges. Anyone at the centre would see galaxies falling inwards, pulled by the region's high gravity, but Ellis then used Einstein's theory of relativity to imagine space curving in on itself in such a way that it created a second centre (or 'anti-centre') at the low-density edges. In his diagrams, this resulted in an egg-shaped universe, and, for our purposes, that image will suffice. Crucially, anyone living at the top of the egg (the anti-centre) would see galaxies rushing away in all directions, falling towards the bottom, high-gravity centre.

Ellis's egg universe neither expanded nor contracted. It had no beginning in time, nor would it have an end. It had always existed in the same form and always would, because the singularity at the bottom acted as a kind of cosmic recycling centre, consuming old matter from burned-out galaxies and spewing it back out as hot, fresh hydrogen that eventually formed into new galaxies.

Ellis placed our galaxy at the very top of the egg, in the

low-density anti-centre. As such, this was an identifiable, unique location in the universe, but he emphasized that being at this spot didn't mean that we, as a species, were somehow special. To the contrary, the anti-centre just happened to be the only place in this universe cool enough for life to survive. It was the only place where we could possibly be.

This universe Ellis had dreamed up was extremely odd – he acknowledged that – but his point was that it could very well be the universe we actually inhabit. After all, the cosmic view from the top of the egg would be exactly the same as the view seen if we were living on a dot in the inflating-balloon universe of standard cosmology. In each case, we would see galaxies moving away from us on all sides. In the inflating-balloon universe, this effect would be caused by the expansion of the entire universe, while in the egg universe it would be caused by galaxies being pulled towards the high-gravity singularity. But, from our location, there would be no way to tell the two apart. Both models would produce identical observational data.

Is that it? Do we live at the top of a giant egg? If so, this would completely upend modern cosmology. After all, Ellis's model summarily dispensed with the Big Bang, imagining the universe to have no beginning. Astrophysicists reacted to this with incredulity. Reviewing it in *Nature*, the physicist Paul Davies joked that it was lucky for Ellis that burning at the stake for heresy had gone out of fashion.

Of course, the egg universe hasn't brought down modern cosmology. As Ellis continued to tinker with his model, he decided that he couldn't get it to work to his satisfaction with Einstein's general theory of relativity, and this led him to abandon the idea. So much for the egg universe.

But this didn't alter the bigger point he was trying to make, which remained as relevant as ever. This was the problem of the large-scale structure of the universe – that it has the potential to trick us, undermining our attempts to make sense of the cosmos in which we live.

Based on the astronomical observations that we have, the standard, inflating-balloon model of the universe is, in fact, a logical conclusion to arrive at. It's entirely reasonable to assume that, if we see galaxies receding from us on all sides, the same phenomenon is occurring everywhere else in the universe. This is the simplest assumption to make.

The problem is, just because this assumption is simple and reasonable doesn't mean it's correct. The universe never gave any guarantee that it was going to make things easy for us. And we can never zoom out from our current position to get a grand view of the whole universe in order to confirm that the inflating-balloon model actually is correct. So, the possibility always remains that some other, weirder, more complex large-scale way in which the cosmos is structured could be producing all the effects we observe. *

After abandoning the egg model, Ellis proceeded to make it something of a hobby to dream up alternative cosmic structures of this kind. One of his ideas is that we may live in a small universe. Astronomers believe the universe is infinite in size, but Ellis suggested that maybe it's only several hundred million light years across, but it curves around so that, if you were able to travel that entire distance, you'd end up back where you started. This would mean there aren't as many galaxies surrounding us as we think; it's just an optical illusion. We're seeing the same patch of galaxies over and over, as if we're staring into two parallel mirrors producing an endless series of reflections. Ellis happily concedes that this probably isn't how the universe really is, but there's no way to know for sure.

Another of his ideas is that while the Earth may not be at the centre of the entire universe, it may be at the centre of a 'cosmic void'. This was suggested to him by the discovery, made in the late

---

* See 'What if our universe is actually a computer simulation?' for one such possibility.

1990s, that the expansion of the universe is accelerating. To account for this, astronomers have theorized that a mysterious 'dark energy' is causing the acceleration, but Ellis proposed instead that the distribution of matter in the universe may be uneven. In most of it, he said, the density of matter is very high, but there are bubbles of space, 'cosmic voids', in which it is low. Our galaxy might be floating in the centre of one of these low-density voids, and the galaxies trapped in this bubble with us are all falling outwards, pulled towards the surrounding regions of higher gravity at an ever-increasing rate. If this is the case, the apparent accelerating expansion of the universe would only be a phenomenon local to our region of space.

Again, Ellis doesn't insist that this is the way the universe really is. It's more of a philosophical point he's trying to make about the nature of cosmology itself. It's a curious discipline, because cosmologists study an object (the cosmos) the vast majority of which they can never see or interact with. It's a bit like trying to figure out what a building looks like when all you've got to go on is one brick. The best you can do is to assume that the entire structure is similar to that one brick and proceed on that basis. But what if the brick isn't actually representative of the building as a whole? What if it was part of a facade? You'd have no way of knowing.

That's what it's like for cosmologists. They try to figure out what they can about the nature of the universe based on the assumption that the part of it we can see is representative of the whole thing. But there's no way to be certain. There's an absolute, hard limit on our knowledge, forcing us to live with the nagging suspicion that the universe might actually have some kind of large-scale structure completely different from anything we could ever imagine.

What if planets can explode?

If you're the type that's prone to worry, the natural world offers a lot to fret about: hurricanes, earthquakes, tsunamis, super-volcanoes, asteroid impacts and any number of other threats could spell your doom. But what about exploding planets? What if Mars suddenly exploded and showered us with meteors? Or, even worse, what if the Earth itself burst apart without warning in a fiery cataclysm? Could this happen? Should you add the possibility to your list of concerns?

Short answer: no. Geologists feel quite sure that planets, left to their own devices, don't blow up. Advocates of the exploded-planet hypothesis, however, beg to differ. They argue that planets do have a disconcerting tendency to abruptly and catastrophically detonate, and that there's evidence this has already occurred multiple times within our own solar system.

The genesis of the exploded-planet hypothesis traces back to the mid-eighteenth century, when an astronomical puzzle emerged: the mystery of the missing planet. A German professor of mathematics, Johann Titius, noticed something unusual about the positions of the planets in our solar system. Their distances from the sun seemed to follow an orderly pattern. Each planet lay approximately twice the distance from the sun as its inner neighbour. Titius had no idea *why* the planets would be spaced out like that; he just noted that they were – except for one conspicuous exception. There was an extremely large, empty gap between Mars and Jupiter, and, according to the pattern, there should have been a planet in that space. He recorded this observation as a footnote in a book he was translating from French into German.

This wasn't exactly an attention-grabbing way to announce a new discovery. The footnote could easily have gone unnoticed, but the German astronomer Johann Bode happened to see it and decided Titius might have stumbled upon a law that regulated planetary distances. So, Bode included the idea in an astronomy textbook he was writing.

That might have been the end of the story. Titius's observation didn't otherwise attract much attention, until 1781, when something happened that made it the focus of intense scrutiny. In that year, the British astronomer William Herschel discovered the planet Uranus, out beyond Saturn. This was the first time a new planet had ever been found – not counting the discovery by prehistoric sky-watchers of Mercury, Venus, Mars, Jupiter and Saturn – and, to everyone's surprise, Uranus lay almost exactly where Titius's pattern predicted it should, as a seventh planet. Astronomers at the time didn't think this could be mere chance – they thought Titius and Bode had to be on to something and that the pattern really was a law. It came to be known as Bode's Law, which was a little unfair, since Titius was the one who had actually figured it out. Bode had merely publicized it better!

The acceptance of Titius's pattern as a fully fledged law immediately focused attention on that gap between Mars and Jupiter. If the pattern had been right about the position of Uranus, where was the planet in the Mars–Jupiter gap? An intensive search promptly began among astronomers throughout Europe, with the result that, in the early nineteenth century, something was found at exactly the right spot. But it wasn't a planet. It was a bunch of asteroids. They were, in fact, the first asteroids ever found, and this region of the solar system became known as the asteroid belt.

If you go looking for a house, the position of which is indicated on a map, but when you get there all you find is a pile of rubble, it would be reasonable to conclude that the house must have been destroyed. Likewise, if you go looking for a planet, but all you find in its place is a cluster of large rocks, you might suspect that the planet had come to a bad end. That's exactly what the astronomer Heinrich Olbers concluded. In 1812, he proposed that the asteroid belt between Mars and Jupiter must be the broken-up remains of a former planet that had either exploded or had been destroyed in a collision. In this way, the unsettling idea that planets may not be as stable as they seem, that sometimes they simply blow up of their own accord, entered the astronomical imagination.

Olbers' disturbing vision didn't remain scientific dogma for long. Astronomers soon abandoned it in favour of the more reassuring theory that the asteroids were mini proto-planets that had been prevented from combining together into larger bodies due to the interference of Jupiter's massive gravity tugging on them.

Nor did Bode's Law retain the status of a law. By tradition, it still bears that name, but the planet Neptune, discovered in 1846, beyond Uranus, was way out of sync with the pattern. So, astronomers concluded that there was no law of planetary distances, and that the pattern observed by Titius was mere coincidence.

But it was only a matter of time before the catastrophist view of nature reasserted itself. In 1972, the British astronomer Michael Ovenden published an article in *Nature* in which he revived Olbers' spectre of a blown-up planet between Mars and Jupiter. Adding more drama to the story, he envisioned this vanished world to have been a massive gas giant, ninety times the size of Earth. He named it Krypton, after Superman's exploded home world.

Like Titius, Ovenden suspected there had to be an underlying order to the positions of the planets. He theorized that, over time, the planets will always settle into positions at which they have the minimum gravitational interaction with each other. He called this his principle of planetary claustrophobia. But what puzzled him was why the planets weren't all in these locations. Specifically,

Mars was relatively close to Earth, but far away from Jupiter. According to his calculations, it should have been closer to Jupiter, and this is what led him back to the idea of a missing planet.

He proposed that the current positions of the planets would make sense if a massive planet had once existed between Mars and Jupiter, but that it had abruptly 'dissipated' about sixteen million years ago. Following this disappearing act, Mars would have started slowly inching closer to Jupiter, though it would still take millions of years before it reached the position of minimum interaction.

What, however, could have made a planet dissipate so quickly? Ovenden saw only one possibility: it must have exploded. The bulk of the planet, he believed, must have been swallowed up by Jupiter, leaving what remained to become the asteroid belt.

Ovenden's hypothesis inspired a show titled 'Whatever Happened to Krypton?' that toured North American planetariums in the late 1970s, but it didn't particularly impress astronomers. Most of them ignored it, with the significant exception of one young researcher who was profoundly taken with the idea. This was Tom Van Flandern, who at the time was working at the US Naval Observatory. Up until then, Van Flandern had been an astronomer in good standing, who didn't stray far from conventional wisdom, but the idea of spontaneously detonating planets deeply appealed to him. Under the influence of Ovenden's hypothesis, he took a sudden, sharp turn towards scientific unorthodoxy. Over the following decades, to the bafflement of his colleagues, he transformed himself into a kind of weird astronomy celebrity, frequently expounding his against-the-mainstream theories (such as his claim that geological features on Mars show the handiwork of an extraterrestrial intelligence) on late-night radio shows.

As he pondered Ovenden's catastrophist vision, Van Flandern concluded that there was more evidence, beyond the existence of the asteroid belt, of a long-ago planetary explosion. In fact, he came to believe that the solar system was a landscape scarred by the incendiary violence of its past, with a variety of solar-system oddities suddenly acquiring new significance in his eyes.

There was, for example, the curious pattern of cratering on Mars. The planet's southern hemisphere is heavily cratered, whereas its northern hemisphere is relatively smooth. The conventional explanation for this 'Martian dichotomy' (as it's called) is that it's the result of a massive asteroid impact early in the planet's history that created a magma ocean, resurfacing and smoothing out the northern half. Van Flandern instead argued that Mars must originally have been a moon of Ovenden's exploded planet. So, when the planet blew up, the hemisphere of Mars facing towards it was hit, full force, by the blast, cratering it, while the opposite side remained unscathed.

Then there was the unusual colouration of Saturn's moon, Iapetus. One half of it is dark, while the other half is bright white. Mainstream theory attributes this two-tone effect to differing temperatures on its surface, which produce an eccentric pattern of water-ice evaporation. The ice accumulates on the side that's coldest, rather than where it's relatively warm. (Although, the entire moon, by Earth standards, is freezing cold.) Van Flandern, however, argued that the dark half came to be that way when it was blackened by the blast wave generated by the exploding planet. Since Iapetus rotates extremely slowly, only one side of the moon ever faced the blast wave, which would explain, he said, why none of the other Saturnian moons show similar colouration – they all spun fast enough to be uniformly blackened.

As the years passed, Van Flandern's vision of the violent past of our solar system grew ever more elaborate. He concluded that exploding planets hadn't just been a one-time event in its history, but were a recurring feature. Initially, he upped the explosion count to two planets, which he designated by the letters V and K (K for Krypton, in a nod to Ovenden). His reasoning was that there were two distinct types of asteroids in the asteroid belt, so there must have been two planets. But, by the end of his life, in 2009, he had worked his way up to a full six exploded planets: two in the asteroid belt between Mars and Jupiter, another two to account for the Oort cloud of comets beyond Neptune, and a final two, just for good measure. Planets were apparently going off like fireworks in our solar system.

Of course, it doesn't matter much how scarred and rubble-strewn our solar system might be. If planets can't explode, they can't have been the cause of the destruction. And, really, why would they explode? They're giant lumps of rock and gas. There's no obvious reason for them ever to detonate spontaneously.

Van Flandern realized this, so he and his fans set to work to identify the hidden explosive force that could transform planets into ticking time-bombs. Among the ideas they came up with were gravitational anomalies, antimatter beams from the centre of the galaxy, and even interplanetary war. It wasn't until the 1990s that a more plausible possibility swam into view – conceivable because it didn't involve any impossible physics, not because it in any way reflected orthodox thinking in science. It was the georeactor hypothesis.

The idea was that the cores of some planets might consist of massive balls of the highly radioactive element uranium, acting as natural nuclear fission reactors, or georeactors. Some isotopes of uranium are fissile, which means that when they absorb a neutron, they split in two, releasing energy and more free neutrons. If you collect enough uranium together, the fission of one atom will trigger the same in its neighbour, which will do likewise for its neighbour and so on. The process becomes self-sustaining, producing enormous amounts of heat and energy. This is how man-made nuclear reactors work.

So, imagine a lump of uranium five or ten miles wide, cooking away inside a planet. Under normal circumstances, uranium reactors just release energy; they don't explode. Something has to compress the uranium into a tight ball, crowding the atoms together and causing them to reach a supercritical mass to produce a detonation. Nuclear bombs achieve this through the use of conventional explosives. A planetary georeactor, therefore, would remain stable as long as it was left alone. But there are things that could potentially set it off. The shockwave of a very large asteroid hitting the planet might do the trick, and, if something like this did happen, the resulting nuclear blast would absolutely have enough force to rip apart a planet, sending chunks of it flying clear out of the solar system.

The idea that natural nuclear reactors might exist at the cores of planets wasn't Van Flandern's idea. He simply recognized, as soon as he heard about it, that it could provide a mechanism for planetary destruction. The hypothesis was the invention of J. Marvin Herndon, a mining consultant with a doctorate in nuclear chemistry, who hit upon it in an attempt to answer an entirely different question: why Jupiter, Saturn and Neptune all radiate far more heat than they receive from the sun.

The standard answer is that these planets are still radiating heat left over from their formation, but Herndon thought they should already have cooled down – especially since, being mostly gas, they lack the insulation to trap in their primordial heat. He was mulling over this mystery while standing in line at the grocery store during the early 1990s when he had a sudden epiphany. He remembered that, back in 1972, French scientists in Gabon had discovered an underground pocket of uranium that, they realized upon analysis, had been acting as a natural nuclear reactor some two billion years ago, before exhausting its fuel. Similar naturally formed georeactors have been found subsequently. Their existence had actually been predicted in 1956 by the physicist Paul Kazuo Kuroda, but, at the time, the scientific community had scoffed at his idea. He had difficulty even getting it published – another example of a weird theory that became true!

So, Herndon reasoned, if it's possible for a natural nuclear reactor to form in the crust of the Earth, as it evidently is, it might also be possible for one to form at the core of a planet. There's plenty of uranium around, and it's the heaviest metal commonly found in nature. Under the right conditions, during the process of planetary formation, it might sink directly into the core and concentrate there, triggering the fission process. If this had happened in the case of the outer planets, it could certainly explain the excess energy radiating from them.

Herndon originally applied this reasoning only to Jupiter, Saturn and Neptune, but he soon extended it to the Earth as well, making the case in several articles published in the *Proceedings of the National Academy of Sciences*. He pointed out that the Earth produces enough energy to power a strong magnetic field that shields us from the worst effects of the solar wind. Where, he asked, does all this energy come from? The conventional answer is that it's a combination of residual heat, radioactive decay and gravitational potential energy, but Herndon doubted these sources would suffice. A georeactor, on the other hand, would easily power the magnetic field. Daniel Hollenbach, a nuclear engineer at Oak Ridge National Laboratory, used computer simulations to help Herndon confirm that this, in principle, was possible.

This brought the exploded-planet hypothesis right to our doorstep. Herndon himself didn't make the connection that a georeactor might detonate, but others did. If he was right that there's a five-mile-wide ball of burning-hot uranium at the core of the Earth, then it might only be a matter of time before our home world becomes the next Planet Krypton.

But, forget about the possibility that the Earth *might* explode – what if it already has? This idea forms the most recent and arguably most sensational development in the ongoing saga of the exploded-planet hypothesis. Obviously, the explosion couldn't have been big enough to completely demolish the Earth – we're still here. But it could have been enough to leave some dramatic evidence behind, which now floats above our heads most nights: the moon.

The origin of the moon is a genuine mystery to scientists, which may seem paradoxical given what a familiar object it is. You'd think they would have figured out where it came from by now.

And yet, explaining how the moon formed has proven to be extremely difficult.

The problem is that the moon is very large – far too large for the Earth's gravity to have captured it if it happened to have been speeding by. Moon rocks are also chemically almost identical to the rocks on Earth. It's as if a giant ice-cream scoop took a chunk out of the Earth's mantle and placed it in orbit, up in the sky. The challenge for scientists has been to explain how that scoop got up there.

The leading theory, proposed in the mid-1970s by the planetary scientists William Hartmann and Donald Davis, imagines that a Mars-sized object, nicknamed Theia by astronomers, struck the Earth at an angle, causing a huge chunk of the mantle to splash up into space, where it initially formed a ring around the Earth before eventually congealing into the moon. The problem with this theory, as planetary scientists acknowledge, is that some chemical signatures of Theia should have remained in the lunar rocks, but they don't seem to be there. Which brings us straight back to the exploded-planet hypothesis.

In 2010, the Dutch scientists Rob de Meijer and Wim van Westrenen proposed that the moon could have formed by being blasted out of the side of the Earth. They hypothesized that, while the Earth was still very young, a large concentration of uranium formed near its core. An asteroid impact then might have detonated this uranium, launching a massive chunk of the Earth's mantle into space, where it became the moon. This would explain the chemical similarity between the Earth and the moon, because they were once one and the same.

Van Flandern didn't live long enough to learn of this newest wrinkle to the exploded-planet hypothesis, but he doubtless would have appreciated it.

Asteroid rubble where a planet should be, a scarred solar system, georeactors, the moon overhead . . . These are the clues that lead like a trail of breadcrumbs to the conclusion that our planet has the potential to suddenly go boom. But, before you rush out to buy exploding-planet insurance, rest assured that orthodox science sees no reason to worry. As the saying goes, keep calm and carry on!

For a start, astronomers are pretty sure that no planet has ever exploded in the solar system. They point out there's actually not that much stuff in the asteroid belt between Mars and Jupiter. If you lumped it all together, its total mass would only be a few one-hundredths of the mass of our moon, which hardly seems enough to be the remains of a planet, even if much of it did disappear into Jupiter.

Geologists, likewise, seriously doubt that georeactors could ever form at the core of planets. It's true that uranium is very heavy and there's plenty of it around, but uranium likes to bond with lighter elements, particularly oxygen, which should prevent it from ever sinking to a planet's core. We're probably safe.

However, if one were inclined to indulge paranoia, such assurances might not completely satisfy. After all, saying that uranium *should* bond with oxygen is not the same as saying that there's absolutely no way for a planetary georeactor to ever occur. Nothing in physics or geology explicitly forbids it. What if a planet formed from materials low in oxygen? Then perhaps it could happen. At least, it might be vaguely within the realm of possibility – and, with it, the chance of explosive apocalypse.

The exploded-planet hypothesis also recalls a centuries-old dispute within Earth science between so-called catastrophists and uniformitarians. The latter argue that a good geological theory should be based on the identification of natural processes ongoing in the present, such as erosion and sedimentation. The theory will then extrapolate backwards, assuming that these processes have been occurring in a uniform, consistent manner over time. The catastrophists, on the other hand, maintain that sometimes weird stuff happens that doesn't have an exact parallel in the present. On occasion, cataclysmic events shake up the whole system and the pieces fall back down in an entirely new order.

Uniformitarianism traces back to the work of the late-eighteenth-century Scottish geologist James Hutton, regarded as the founder of modern geology. The cause of catastrophism, however, was first taken up by Biblical literalists, who insisted that Noah's flood had produced the rock formations we find today. As a result, it gained a bad reputation within the scientific community. Uniformitarianism came to be regarded as the hallmark of proper Earth science.

But, since the mid-twentieth century, catastrophism has been making a comeback. Researchers began to realize that rare events such as super-volcanoes and massive meteor impacts have had a profound effect on the history of life and the Earth. The recognition, in the 1980s, that an asteroid strike was the most likely cause of the extinction of the dinosaurs, marked a turning point in the rehabilitation of catastrophism. A strong tendency remains among scientists, however, to be sceptical of explanations that invoke catastrophes, and one can't help but wonder if the exploded-planet hypothesis suffers on account of this old suspicion.

As we grow increasingly aware of the many extravagant terrors that the universe is capable of unleashing upon us – rogue black holes that could materialize in the middle of the solar system and swallow us whole, or stray gamma bursts from distant galaxies that could abruptly incinerate us – perhaps the idea of exploding planets will come to seem less outrageous. If the phenomenon is possible, it would, after all, merely be one more item in nature's awe-inspiring arsenal of destruction.

Weird became true: the heliocentric theory

In 1543, the Polish priest Nicolaus Copernicus published *De Revolutionibus Orbium Coelestium* (*On the Revolutions of the Celestial Spheres*), in which he offered what is arguably the greatest weird theory of all time. His outrageous claim was that the Earth revolved around the sun. This contradicted what almost every astronomer since the beginnings of astronomy had believed to be true, which was that the Earth lay at the centre of the universe and everything else – the sun, moon and planets – revolved around it.

Nowadays, it may be difficult to think of Copernicus's theory as being odd. After all, we know beyond a shadow of a doubt that the Earth does indeed revolve around the sun. From our perspective, his theory is the epitome of wisdom, which is why he's popularly celebrated as a great thinker who managed to part the veils of superstition that had shrouded knowledge up until the sixteenth century. But, given what was known about the natural world in the 1540s, his theory was incredibly weird.

Let's consider why a geocentric, or Earth-centred, model made so much sense to ancient and medieval people. For a start, it aligned with the evidence of their own eyes. Anyone could look up and see that the sun, moon and planets travelled from one corner of the sky to the other. The heavenly objects were obviously what was moving, not the Earth.

It also had the weight of tradition and authority behind it. Up until the sixteenth century, no scholar, with the minor exception of the ancient Greek astronomer Aristarchus, had ever seriously questioned the assumption that the Earth was at the centre of the cosmos, and Aristarchus had offered up the idea merely as a speculative hypothesis.

Geocentrism also explained gravity in an intuitive way. Scholars since the time of Aristotle had taught that objects fell down because they had a built-in urge to move towards the centre of the cosmos, which was the Earth. It made sense that things would gravitate towards the middle, rather than towards a random point on the edge.

Most importantly though, it worked. In the second century AD, the great Egyptian astronomer Claudius Ptolemy had devised a mathematical model of the solar system, based upon geocentrism. His system was quite complicated because he had been forced to introduce some creative geometry to explain why the planets occasionally appeared to move backwards in the sky. Nowadays, we know that this 'retrograde' motion occurs because we're watching the planets orbit the sun as we ourselves simultaneously orbit it, but, to explain this motion from a geocentric perspective, Ptolemy had concluded that the planets circled the Earth while simultaneously circling around the path of their own orbit, in what is known as an epicycle. This meant that his system had all kinds of things spinning around each other, like gears turning around gears turning around other gears in a complex machine. But it did nevertheless accurately predict the movement of the planets, which seemed to indicate that it represented the way the cosmos really was.

Then, almost 1,500 years later, along came Copernicus, who proposed tossing the geocentric cosmos overboard and replacing it with a heliocentric, or sun-centred, system. Why? Not because he had any new observational evidence about the movement of the planets. He didn't. His entire argument rested upon the fact that he had devised a new mathematical system which, he claimed, could predict the movement of the planets as well as the Ptolemaic system.

Copernicus had hoped his system would have no need for the complex epicycles and therefore would be simpler, but it didn't turn out that way because he made one crucial mistake. He assumed the Earth revolved around the sun in a perfect circle, which it doesn't. Its path actually takes the form of an ellipse – more like an egg shape. As a result of this error, he had to include epicycles anyway to make his predictions align with observed data. This made his system just as mathematically convoluted as the Ptolemaic one.

In addition, it actually had some stark disadvantages when compared to the geocentric model. By moving the Earth away from the centre of the cosmos, he lost the explanation for gravity. In fact, if the sun was actually at the centre, it wasn't clear why the Earth didn't plunge downwards into it. Copernicus had no answer to this puzzle.

Even more disturbingly for his contemporaries, the heliocentric system put the Earth in motion. If Copernicus was to be believed, we were all living on a giant spinning rock, hurtling through space. It didn't feel like the Earth was moving. If it was, his critics asked, why wasn't there a constant headwind blowing from the direction of its travel? (At the time, no one knew that space was a vacuum.) Why wasn't everything on its surface flung off into space? Again, Copernicus had no answer to these troubling questions.

With all these problems, most scholars at the time naturally rejected his heliocentric model. Any reasonable person *should* have rejected it, based upon the slim evidence Copernicus provided. It was only in the seventeenth century that other scholars fixed the shortcomings of his theory. This began when the Dutch spectacle-maker Hans Lippershey invented the first telescope in 1608, which promptly inspired the Italian scholar Galileo Galilei to build his own. With it, he saw moons around Jupiter, which proved that not everything was orbiting the Earth. Soon after, the German mathematician Johannes Kepler showed that the dreaded epicycles could be eliminated from the Copernican system if the planets moved in elliptical orbits rather than perfect circles. And finally, in the 1680s, the Englishman Isaac Newton tied everything together by developing the theory of gravity. This explained why apples fall to the ground and also why the planets move in ellipses – they're constantly falling towards the sun and missing it. The accumulation of all this new evidence convinced scholars of the validity of the heliocentric theory.

But, given the weirdness of what Copernicus had suggested (in the context of sixteenth-century knowledge) and the weak arguments he presented for it, one has to wonder what exactly inspired him to come up with his theory in the first place. What was he thinking? This question has long puzzled historians.

One idea is that he may have been trying to enforce a belief that heavenly objects move in perfect circles. Both ancient and medieval scholars took it for granted that celestial objects naturally moved in circles rather than straight lines because, they felt, the circle was a perfect shape and therefore was appropriate for the heavenly sphere. The Ptolemaic model took a few minor liberties with this principle because the sheer complexity of the system made having to tweak the rules in various ways inevitable. Copernicus may have decided that this made it irredeemably corrupt. It's certainly true he rigidly enforced the principle of circular heavenly motion in his own system, which, as we've seen, is why he ended up having to fall back on epicycles.

A more controversial idea, put forward recently by the University of California historian Robert Westman, is that Copernicus was trying to defend and improve astrology, the study of how the planets supposedly influence the health and fortunes of people on Earth. The validity of astrology was something that, again, both ancient and medieval scholars took for granted. But, in the late fifteenth century, the Italian philosopher Pico della Mirandola had attacked it as an inexact science. One issue he singled out was the difficulty of knowing the exact positions of Mercury and Venus with respect to the sun in the Ptolemaic system. If astronomers didn't know this, he asked, how could astrologers possibly make accurate forecasts?

Copernicus solved this problem in his heliocentric system. If everything was orbiting the sun, the positions of the planets became unambiguous. The faster a planet orbited the sun, the closer it was to it. Perhaps Copernicus had hoped that this new clarity would lead to more accurate astrological predictions. Westman points out that Copernicus did both live and study with an astrologer while attending university.

Whatever may have been Copernicus's underlying motive for developing the heliocentric theory, it's clear that he wasn't driven by what we could consider today to be good reasons. His beliefs were deeply rooted in medieval assumptions about the nature of the world. In essence, he managed to come up with the correct model of the solar system for entirely incorrect reasons. He lucked out.

Nevertheless, he left behind a powerful legacy of iconoclasm. He showed that it was possible to come up with a weird theory that directly challenged the most basic assumptions about the world and to end up being right. Weird theorists have been trying to emulate his example ever since.

What if our solar system has two suns?

Something in outer space is killing Earthlings. Every twenty-six million years, it slaughters a whole bunch of us. It causes many species to go entirely extinct.

The existence of this extraterrestrial assassin was first detected in the early 1980s by the University of Chicago palaeontologists John Sepkoski and David Raup, who had compiled a huge database of marine fossils found in sedimentary rocks. It was the most comprehensive database of its kind and it allowed them to start examining various large-scale patterns of evolution, such as when families of marine life had gone extinct and how often this had happened.

As they graphed their data, what they found shocked them. There was a distinct periodicity in the rate of mass extinctions. The spikes in their graph were unmistakable. Approximately every twenty-six million years, for the past 250 million years, a whole group of species had abruptly disappeared. They checked and double-checked their data, but the periodicity seemed to be a real phenomenon.

What, they wondered, could have caused such regularly repeating mass extinctions? They couldn't think of any natural phenomenon on Earth that would recur on a twenty-six-million-year cycle. So, when they published their findings in 1983, they suggested that the mass die-offs must have been triggered by something non-terrestrial. There was a cosmic serial killer on the loose.

An astronomical murder mystery immediately caught the attention of scientists. As sleuths took up the case, they quickly decided that, if something coming from space was regularly killing Earth creatures, it was almost certainly one of two things: asteroids or comets. Asteroids are essentially big rocks, whereas comets are lumps of ice, dust and rock. A big enough one of either, if it impacts the Earth, can cause serious death and destruction.

But these would merely be the murder weapon. The more puzzling question was what force might be wielding that weapon. There had to be some astronomical phenomenon that was periodically flinging those objects in our direction. But what was there out in space that exhibited such a regularly repeating pattern over such a vast time scale?

One idea, suggested by Raup and Sepkoski, was that the spiral arms of the Milky Way Galaxy might be the culprit. Our solar system orbits the centre of the Milky Way once every 230 million years, but we're moving slightly faster than the arms of the galaxy rotate. As a result, as we travel along, we move in and out of the arms. It was possible that, whenever we moved into an arm, the slightly higher density of matter there was gravitationally disturbing the orbits of asteroids and comets, causing a bunch of them to fall into the inner solar system, where some of them hit the Earth.

It was an interesting idea, but analysis revealed that the periodicity was all wrong. We only cross into an arm about once every hundred million years. This gave the spiral arms a pretty good alibi; they couldn't be the killer.

NASA scientists Michael Rampino and Richard Stothers offered another idea. They suggested that the perp could be the flat plane of the galaxy. The Milky Way is a gigantic flat disc of matter that spins around. Our solar system moves around with it, but as it does so it simultaneously bobs up and down with a wave-like motion, rising slightly above the surface of the disc, then sinking below it, over and over again. The two researchers argued that, each time our solar system passed through the plane of the galaxy, this might cause a gravitational disturbance that disrupted the orbit of comets, sending them on a collision course with our planet.

The length of the periodicity was approximately right. We pass through the plane of the galaxy every thirty-three million years. But there turned out to be other problems. The matter in the plane of the galaxy is very diffuse. Astronomers found it hard to believe that passing through it would produce much of a disturbance. Also, we're currently in the middle of the galactic plane. If the hypothesis of Rampino and Stothers was correct, we should be experiencing a mass extinction just about now, but, according to the timetable drawn up by Raup and Sepkoski, the next one isn't due for another thirteen million years – luckily for us! The two periodicities, therefore, didn't sync up. Once again, the suspect had an alibi.

With spiral arms and the plane of the galaxy ruled out, University of California physicist Richard Muller stepped forward with an altogether more radical hypothesis. He proposed that our solar system has two suns. There was the familiar one, which we all know and love, but there was also a companion star, an evil twin that was periodically flinging comets at us.

Even people who think they know nothing about astronomy are sure of at least one basic fact: our solar system has one sun. Look up in the sky and there it is, shining away. There are not two of them, as there are on Luke Skywalker's home world of Tatooine. So, to claim that our solar system actually has two suns might seem perverse. And yet, Muller had a reasonable argument to back up his claim, and astronomers were willing to hear him out.

The hypothesis was primarily Muller's brainchild, but he had help working out the details from Marc Davis of Lawrence Berkeley Laboratory and Piet Hut of Princeton University. All three appeared as co-authors on the article detailing the hypothesis that appeared in *Nature* in 1984. They explained that our sun might have a distant companion that circled it in an extremely elongated, elliptical orbit, taking a full twenty-six million years to complete one orbit. At its furthest distance, this companion would be a vast fourteen trillion miles from our sun, but over time it swept in closer, until it was a mere three trillion miles away.

At this distance, it would pass through the Oort cloud, a massive cloud of trillions of comets that surrounds our solar system at its furthest edge, and each time it did so it would dislodge billions of comets from their orbits, sending them falling into the solar system. A few of those billions would inevitably end up hitting the Earth. After having wreaked this havoc, the death star would then retreat back into the depths of space, on its long arcing journey, to return again in another twenty-six million years. This cycle, the three authors suggested, had been repeating for hundreds of millions of years.

As an explanation of the recurring mass extinctions, this hypothesis worked. There were no problems of non-syncing periodicities that could possibly give the death star an alibi. More significantly, what other explanation could there be? Astronomers were running out of things in the universe that could plausibly be hurling comets at us every twenty-six million years.

The authors also noted that the majority of known stars, over two thirds of them, are thought to have companions. So, statistically, it was more probable than not that our sun was part of a binary system. It was true that our sun's companion would need to have a highly eccentric orbit in order to allow it to pass through the Oort cloud only once every twenty-six million years, but that didn't make it impossible – just slightly odd.

The three authors suggested naming our sun's hypothetical companion Nemesis, after the Greek goddess of vengeance. And, if that name didn't work out, they wrote, George might work as an alternative. This was apparently an attempt at scientific humour. However, George didn't make it through the editorial process. The editor at *Nature* made an executive decision and chose Nemesis.

There was just one problem with their hypothesis, which they noted with a touch of understatement: 'The major difficulty with our model is the apparent absence of an obvious companion to the Sun.'

That *was* an important detail! If our solar system had a second sun, you would think that someone might have noticed it by now. But not necessarily, they argued. Nemesis could be a red-dwarf star. These are the most common type of star in the galaxy, but they're small and very dim, a mere fraction of the size of our more familiar sun. This would explain why it had never been seen. It had got lost in the background, blending in with all the other stars. Although, now that the possibility of its existence had been realized, the hunt for it could begin.

The arguments for Nemesis were perfectly legitimate. If this second sun existed, it would neatly explain the puzzling periodic repetition of mass extinctions. So, the scientific community duly took the hypothesis under consideration. There was, nevertheless, a bit of eye rolling among sceptics. This wasn't just because of general wariness at the idea of a death star; it was because they had heard similar claims before. By the 1980s, there was already a firmly established but slightly fringe tradition in astronomy of suspecting that some kind of massive object, often referred to as Planet X, lay undiscovered at the edge of the solar system.

There's a definite mystique about searching for hidden things. It stirs the imagination, and quite a few disciplines feature this kind of hunt. Biology has an active subculture of cryptozoologists, who are absolutely convinced that nature is full of creatures yet to be found, the most famous of these elusive critters being Bigfoot and the Loch Ness Monster. Archaeology similarly boasts a long history of explorers consumed by quests for lost cities, such as El Dorado. These hunts can easily acquire a fanatical, obsessive tinge, and the search for Planet X was no different.

The astronomer William Herschel planted the initial seed of Planet X back in 1781, when he discovered Uranus. Until then, it hadn't occurred to anyone that there might be planets yet to discover in our solar system. Everyone had assumed that its roll call was complete. With the realization that it might not be, the hunt was on.

The search soon bore spectacular fruit when the French astronomer Urbain Le Verrier, led by irregularities in Uranus's orbit, found the planet Neptune in 1846. But, instead of satisfying the appetite for lost planets, his discovery only amplified it. Planets, moons and asteroids were all soon being examined for irregularities in their orbits. If any were found, these were claimed to be evidence of the existence of yet another planet.

The wealthy American businessman-astronomer Percival Lowell coined the term Planet X in the early twentieth century. It was his name for a massive planet that he believed lay beyond Neptune. He spent the last decade of his life trying to find it and, though he died unsuccessful, the astronomer Clyde Tombaugh took up his quest and, in 1930, found Pluto. However, as scientists became aware that Pluto was a mere dwarf planet, not even as big as our moon, the Planet X enthusiasts grew restless. That wasn't the Jupiter-sized behemoth they were looking for. So, the hunt continued.

With this history in mind, when Muller proposed the Nemesis hypothesis in 1984, sceptics couldn't help but wonder if it wasn't the latest and greatest incarnation of the hunt for Planet X. Except, now, instead of being a mere planet, the mystery missing object had swelled into a sun.

But just because Planet X had achieved some notoriety, that didn't mean it wasn't actually out there. Astronomers conceded this. The same was true for Nemesis. It could exist. The only way to know for sure was to look for it. This, however, was easier said than done. It was a bit like looking for a tiny needle in a haystack the size of Texas, and to do so in extremely dim lighting.

The best hope was that Nemesis might be detected by one of the all-sky surveys that are periodically conducted, in which astronomers use a high-powered telescope to systematically search for and catalogue every visible object in the sky. As luck would have it, in the mid-1980s NASA launched its Infrared Astronomical Satellite (or IRAS) survey, capable of detecting extremely faint objects, but it didn't find Nemesis. Nor did the even more sensitive Two Micron All-Sky Survey (2MASS), conducted from 1997 to 2001. When NASA's Wide-field Infrared Survey Explorer (WISE) space telescope launched in 2009, many astronomers viewed it as the final chance for Nemesis. When this too failed to find it, the general consensus was that this meant our sun doesn't have a companion star.

This has been the central problem for the hypothesis. Absence of evidence isn't necessarily evidence of absence, but the more time has passed without anyone finding Nemesis, the less willing astronomers have been to believe that it exists. And most of them were pretty sceptical about it from the start.

There's been another problem for the hypothesis. In 2010, a pair of researchers re-examined the fossil evidence to make sure that the periodicity found by Raup and Sepkoski really existed. They concluded that it did. In fact, using a larger fossil database, they were able to extend the periodicity back over 500 million years, slightly revising it to a twenty-seven-million- rather than a twenty-six-million-year interval. This might sound like confirmation of Nemesis, but instead they argued that their findings actually suggested a companion star *couldn't be* the cause of the extinctions. Their reasoning was that an object such as Nemesis, with a very large orbit, would inevitably be affected by the gravity of passing stars and the galactic tidal field. This would cause variations in its orbital period, preventing it from maintaining clockwork periodicity. Since these variations weren't seen in the fossil record, Nemesis must not have been the culprit.

Despite these setbacks, and after all these years, Muller hasn't given up hope that Nemesis will eventually be found. He discounts the argument about the regularity of the extinctions, believing that the fossil record is ambiguous enough to allow for the kind of variability that the orbit of Nemesis should display. And, anyway, something has to be causing the extinctions. If it isn't a second sun, what is it? This remains an unanswered question.

Muller has pinned his hopes on the Large Synoptic Survey Telescope, which is currently under construction in Chile and should start full operations in 2022. It's been designed to have a very wide field of view, which will allow huge swathes of the sky to be examined. He says that if that doesn't find anything, then maybe he'll start to question whether Nemesis exists. Or maybe not. Space is vast enough for a star to hide. No matter how carefully you look, the possibility remains that it might be lurking somewhere out there, undetected, after all.

Weird became true: continental drift

Today, geologists accept it as a matter of fact that the continents are constantly moving around the globe at a rate of a few centimetres a year. Although slow, this translates into journeys of thousands of miles over millions of years. But when the young German meteorologist Alfred Wegener first proposed this idea in 1912, he faced a stone wall of resistance from geologists who were firmly committed to the belief that the continents are permanently fixed in place. They didn't just reject his idea, they contemptuously dismissed it as 'Germanic pseudoscience'. It took almost half a century before they finally admitted he had been right all along.

The idea that Africa, the Americas and the rest of the major land masses might be slowly wandering around the globe first formed in Wegener's mind around 1910, when he was a young professor of meteorology at the University of Marburg. He had been admiring a friend's new atlas, leafing through its pages, when he was struck by how the continents seemed to fit together like pieces of a jigsaw puzzle. This was particularly true for the Atlantic coasts of South America and Africa.

Others had noticed this fit before, as early as 1596, when Abraham Ortelius, the creator of the first modern atlas, remarked that the Americas looked like they had been somehow torn away from Europe and Africa. But Wegener was the first to develop the observation into a full-blown theory of geological change. He decided that the jigsaw-puzzle fit meant that the continents must once have been joined together as an enormous supercontinent, and that over the course of millions of years the land masses had drifted apart until they reached their current location.

He detailed this hypothesis in his book *Die Entstehung der Kontinente und Ozeane* (*The Origin of Continents and Oceans*), published in 1915. The timing of this with the onset of World War I wasn't an accident. As a reserve officer in the German army, Wegener had been sent to the Western Front, where he got shot twice, and he wrote the book while recuperating. English speakers first learned of it in the early 1920s, with a full translation appearing in 1924.

The apparent fit of the continents was Wegener's first and main argument, but he carefully gathered other evidence. He pointed out that fossils of identical reptiles and ferns had been found on either side of the Atlantic. The same was true of rock formations, and even living species such as earthworms on the different continents were strikingly similar. This only seemed to be possible if the continents had once been connected.

His theory, however, faced one serious problem. He didn't know how this continental drift had happened. He speculated that the centrifugal force of the spinning planet might cause the movement, or perhaps the tidal forces of the sun and moon. But, basically, he had no idea.

When confronted with Wegener's theory, geologists promptly ripped into its lack of a mechanism. The idea of continents chugging around the globe like giant cruise liners, ploughing through the solid rock of the ocean floor, struck them as self-evidently absurd. They pointed out that, even if the continents could somehow do this, the stress of doing so would surely cause them to fracture and break apart, leaving a trail of wreckage in their wake. No such thing was seen on the Earth's surface.

This was a legitimate point. The idea of moving continents was a hard pill to swallow. It was what made Wegener's theory so weird. Even so, it would have been possible to question the lack of a mechanism while still acknowledging that Wegener had gathered evidence that deserved to be taken seriously. Instead, geologists didn't merely object to continental drift, they tried to completely destroy it.

They called into question Wegener's competence, accusing him of being ignorant of the basics of geology. They criticized his methodology, sneering that he was cherry-picking evidence. They even dismissed the jigsaw-puzzle fit of the continents as an illusion, though they had to contradict the evidence of their own senses to do this, since anyone can look at a map and plainly see that they do, in fact, seem to fit.

But perhaps the most telling criticism geologists levelled against Wegener was that his theory contradicted everything they believed to be true about the Earth. In 1928, the geologist Rollin Chamberlin noted that, for Wegener to be correct, almost the entire edifice of geological knowledge constructed in the past century would have to be wrong. The discipline would need to start all over again. To him and most of his colleagues, this seemed absolutely incomprehensible.

To maintain their old beliefs, though, they had to introduce some weird theories of their own. In particular, there was that evidence of the similarity of species on different continents. It implied that these species had somehow been able to travel across the vast oceans, but how could this be? Had the ferns and reptiles built ships?

The answer the geologists conjured up was that the continents had once been connected by 'land bridges', narrow strips of land that criss-crossed the oceans. They imagined species obediently marching across these bridges at the appropriate times throughout the history of the Earth. These bridges, in turn, had the almost magical ability to rise and fall on command, as if they could be raised from the ocean floor on hydraulic jacks and then lowered back down again when not needed. There was no plausible mechanism for these land bridges, any more than there was for moving continents. And yet conventional theory required them, so they gained the status of geological orthodoxy.

Not all geologists joined the anti-Wegener pile-on. There were some bright spots of resistance to the mainstream, including the South African scientist Alexander du Toit and the esteemed British geologist Arthur Holmes. But the majority firmly rejected continental drift. By the 1950s, the conventional wisdom was that the theory was not only wrong, but that it had been definitively disproven.

Left to their own devices, geologists probably would have continued on forever denying that the continents could move, but advances in other disciplines forced change on them. Sonar technology developed during World War II had led to a revolution in ocean-floor mapping. By the 1960s, this had made it possible to see in detail what lay beneath the water. With this new knowledge, the evidence for continental drift became overwhelming. These maps allowed researchers to see that the fit of the continents significantly improved if one connected them at the continental shelf, this being the actual, geological edge of the continents, as opposed to the coastline, which can change with rising or falling sea levels.

Even more dramatic evidence came from a topographical map of the North Atlantic Ocean floor, produced in 1957 by Bruce Heezen and Marie Tharp. It revealed a massive ridge system running down the centre, like a spine, exactly paralleling the two opposite coasts. All the way down the centre of this ridge system was a rift valley where molten rock was welling up from the mantle. You could almost see in real time the two sides of the ocean floor spreading away from each other, pushing the continents apart. In the 1960s and '70s, Heezen and Tharp completed maps of the remaining ocean floors, showing similar ridge systems running throughout them as well.

As geologists became aware of this new information, it only took a few years for the majority of them to abandon their belief in unmoving continents. Researchers then developed the theory of plate tectonics, which represented an overall vindication of Wegener's concept, though with an important revision. He had imagined only the continents moving, somehow bulldozing through the ocean floors, but plate tectonics envisioned the entire crust of the Earth, including the ocean floors, separated into massive plates that the continents rode on top of. These plates constantly jostled around, spreading apart, sliding past each other, or colliding (in which case, one slowly disappeared, or subducted, beneath the other). All this motion was driven by convection currents deep within the mantle. By 1970, this was firmly established as the new geological orthodoxy. Land bridges and fixed continents were quietly scrubbed from textbooks.

Unfortunately, Wegener didn't live to witness his vindication. He had died long before, in 1930, while leading a research expedition to Greenland. His body remains there to this day, entombed in the ice where his colleagues buried him, now covered by hundreds of feet of snow and drifting slowly westwards with the North American Plate.

What if ten million comets hit the Earth every year?

A lot of stuff falls to Earth from space every year. It's hard to know how much exactly, but scientists estimate that it's as much as 80,000 tons of material. The bulk of this is space dust, but there are also numerous pea-sized meteorites that burn up in the atmosphere. However, in 1986, the University of Iowa physicist Louis Frank suggested that this estimate was too low by many orders of magnitude. To the grand total, he argued, needed to be added ten million 'small comets' that struck the atmosphere every year. Each of these contained, on average, one hundred tons of water. So that added up to one billion tons of water that was falling to Earth annually from space.

To most scientists, this was crazy talk. For a start, they had never heard of such a thing as a 'small comet'. To them, comets were icy objects of fairly significant size, measuring anywhere from one to ten miles across. If one were to hit the Earth, we would know it. If we were lucky, it would detonate with a force equivalent to a nuclear bomb. If we were unlucky, the impact would wipe out our species. And, even if small comets did exist, it seemed wildly implausible that ten million of these things could be striking us annually and somehow no one had noticed until now.

The problem was, Frank hadn't conjured these small comets

out of thin air. He had pictures of them: tens of thousands of satellite images. If these images weren't documenting small comets hitting the atmosphere, what were they showing?

For the first part of his career, until he was in his mid-forties, Frank was a well-respected scientist who adhered to conventional views. He specialized in plasma physics. His accomplishments included making the first measurements of the plasma ring around Saturn and discovering the theta aurora, a polar aurora that looks like the Greek letter theta when viewed from space.

Even after he developed his small-comet hypothesis, he continued his work in plasma physics. As a result, colleagues who knew him from this research often didn't realize he was the same Louis Frank notorious for the small comets. He joked that it was as if there were two of him, like Jekyll and Hyde: 'One appears to be the most conservative of scientists, the other a maverick hellbent on destroying the very foundations of science.'

The events that led to the small-comet hypothesis began in late 1981. One of his students, John Sigwarth, was analysing ultraviolet images of the Earth's atmosphere taken by Dynamics Explorer, a pair of polar-orbiting NASA satellites. He was trying to find evidence of atmospheric ripples that might have been caused by gravity waves, but he kept noticing dark specks in the pictures. There were over 10,000 images, and almost all of them featured at least one or two of these annoying specks.

Sigwarth's first assumption was that something was wrong with the camera's electronics, and he worried that all the images were ruined. He notified Frank, and together they began trying to figure out the cause of the specks. They first tried systematically eliminating any possible technical problems, such as computer glitches, radio transmission noise or failing sensors. At one point, they even suspected it might be paint flecks on the camera. But, one by one, they ruled them all out.

Eventually, they looked at successive picture frames, which led them to discover that the specks could be followed from one frame to another, gradually fading in intensity. The specks also moved mostly in the same direction. This suggested that the specks were an actual phenomenon in the atmosphere, not instrumental error. But what could it be?

Meteors were the obvious candidate, but, although the specks were mere dots, the images displayed huge swathes of the atmosphere almost 2,000 miles wide. This meant that each speck represented something almost thirty miles wide. Any meteor that size would probably cause our extinction.

Instead, they concluded it had to be water vapour. It would take about one hundred tons of vapour to produce specks of the kind they were seeing, but water was the most common substance in the solar system that absorbed ultraviolet light at the wavelength they were observing. It seemed like the logical conclusion.

That much vapour couldn't be coming from the ground; the dots were too high up. Therefore, it had to be coming from space. And that's what finally led to the small-comet hypothesis. Frank concluded that small comets, or 'space snowballs', must constantly be pelting the Earth. He imagined them having the consistency of loosely packed snow. When they hit the atmosphere, they would vaporize and diffuse, producing the dark, UV-light-blocking patches detected by the satellite. To produce this much vapour upon hitting the atmosphere, each one of these comets would need to be about the size of a house.

The way Frank saw it, there was simply no other explanation. But the problem, which he conceded, was the sheer quantity of these things. It was 'an alarming number of objects'. He calculated that about twenty small comets were hitting every minute, which added up to ten million per year. This seemed outrageous, but that's what the satellite data showed. As Sherlock Holmes said, 'when you have eliminated the impossible, whatever remains, however improbable, must be the truth.'

Frank published these results in the April 1986 issue of *Geophysical Research Letters*. He did so even though the journal's editor warned him that making this claim would probably ruin his career. Sure enough, the scientific community reacted with absolute incredulity.

Summoning an ongoing blizzard of space snowballs into existence isn't something to be done lightly. It would add a fundamental new feature to both Earth's environment and space. Scepticism was inevitable, and Frank expected this.

What he found, however, was that two questions in particular seemed to perplex people when they first heard his hypothesis. The question most insistently asked was why these comets weren't destroying satellites or space shuttles. Wouldn't launching anything up into space be like sending it into a firing range? The next most popular query was why, if these things existed in the quantities he claimed, no one had ever seen them before. It sounded like they would be hard to miss. Frank believed he had satisfactory answers to both questions.

He explained that the space shuttles didn't get destroyed because the small comets were extremely fragile, like loosely packed snow, and they began breaking up several thousand miles above the Earth, ripped apart by its gravity and electromagnetic field. By the time they reached near-Earth orbit, where the shuttles flew, about 200 to 300 miles high, they were nothing more than a spray of water, thinner than a London fog. This would cause no damage to a shuttle. In fact, it would be almost undetectable. Frank speculated that shuttles had probably been hit by small comets many times.

He added, however, that if a small comet hit a spacecraft in a higher orbit, it would certainly destroy it. But, since space is big and the comets were small, the odds of an impact were extremely low. Still, there had been cases of spacecraft being lost for unknown reasons. Perhaps small comets had been the culprit.

As for why no one had ever seen these comets, the simple answer was that they were very small and dark. Frank believed they were covered by a thin black crust, like a carbon polymer coating, formed by being bombarded with radiation. This made them almost invisible against the backdrop of the bright stars. Plus, no one had been looking for them.

The slightly more intriguing answer, however, was that perhaps they had been seen, many times, but no one had realized what they were witnessing. Frank suggested they might be the underlying cause behind a variety of unusual phenomena. For instance, astronomers had occasionally reported seeing strange-looking fuzzy or 'nebulous' meteors that appeared to have a diffuse head as they descended through the atmosphere. Perhaps these were small comets.

Then there was the phenomenon of night-shining, or noctilucent, clouds. These are a type of cloud that can be seen in polar regions, where they form in the upper atmosphere, higher than any other clouds. Because of their height, they catch the rays of the setting sun, even after the sky has gone dark, causing them to glow. Scientists aren't sure how water vapour gets high enough to form these clouds, but Frank argued that vapour descending from space could provide an answer.

And what about the anomalous reports of ice falling to Earth from clear skies? This is often blamed on passing aircraft, but such reports predate the invention of manned flight. Frank proposed that these chunks might be the remains of particularly large small comets.

Finally, there were the numerous sightings around the world of UFOs. Frank pointed out that a descending vapour cloud might form a saucer-like shape, which could easily be mistaken for some kind of alien craft. If this was the reality behind UFOs, he mused, it would mean, ironically enough, that they actually were extraterrestrial in origin.

Just in case scientists didn't realize how disruptive these small comets were to conventional wisdom, Frank made sure to spell out some of their more far-reaching implications. He boasted, 'The textbooks in a dozen sciences will have to be rewritten.'

If they were real, then presumably they had been bombarding

the Earth for billions of years, perhaps since it had first formed. In which case, he argued, the comets might have played a role in the development of life by providing the organic material that allowed its emergence.

But that was just the beginning. These space snowballs could also have been the reason for the formation of the oceans. For a long time, their origin was a scientific mystery, but the prevailing belief now is that they formed as a result of the outgassing of molten rocks that pumped steam into the atmosphere which then converted to rain as the planet cooled. Frank insisted that this wouldn't have been a plentiful enough source of water, especially when one considers how much moisture is lost by evaporation to space. His small comets, he said, were the true, never-ending source.

This led to what was perhaps the most startling implication. Frank estimated that the small comets added about an inch of water to the globe every 10,000 years. This had been going on for billions of years, but it was still continuing. Extrapolating into the future, this meant that the Earth would eventually be completely submerged in water. Kevin Costner's film *Waterworld* might have been a prophetic vision.

Waxing more philosophical, Frank argued that the real reason his hypothesis faced so much resistance was because it challenged the prevailing understanding of the relationship between the Earth and the cosmos. Scientists, he said, like to imagine that the Earth is set apart from the rest of the cosmos in splendid isolation. They acknowledge the existence of dangers from outer space, such as the possibility of asteroid strikes or gamma-ray bursts from dying stars, but these all seem comfortably remote. They don't think that astronomy really has anything to do with our daily lives. His small comets, however, suggested that we aren't isolated from the solar system at all. The Earth is being affected by them today, in the here and now, rather than in the distant past or future. Making this claim, he believed, had brought down upon him the wrath of orthodox science.

*

Of course, mainstream scientists didn't agree that they were terrified of contemplating the cosmic connectedness of planet Earth. They just thought Frank's hypothesis was loony, and their primary objection was very basic. It seemed to them that, if small comets existed in the quantities that Frank claimed, their existence should be obvious. You wouldn't need satellite images to intuit their presence.

But they also offered more specific rebuttals. They pointed out that the moon shows no sign of the effect of these comets. If they're hitting the Earth, they must also be hitting the moon. In which case, why isn't the moon covered in water? And why haven't the seismometers left behind by the Apollo astronauts ever picked up any trace of them impacting the lunar surface?

But what about the specks on the satellite images? How to explain those? Most scientists dismissed them as an instrumental glitch. There seemed to be no other explanation.

This touches on an issue at the heart of science: the relationship between knowledge and the instruments that produce it. It was in the seventeenth century that researchers first began to systematically use instruments such as telescopes and microscopes to extend their senses. These instruments quickly became one of the defining features of modern science. They allowed researchers to discover objects that couldn't be seen with the naked eye, but they raised a question: how can you know if what you're seeing is real or an artefact of the instrument? Seeing with your own eyes had always been the gold standard of truth in philosophy, but with the introduction of these instruments, researchers began to claim objects existed that could never be directly seen. You had to trust both the instrument and the skill of its operator.

Sceptics immediately seized on this ambiguity. In 1610, when Galileo used a telescope to discover moons around Jupiter, his critics dismissed what he was seeing as an optical aberration created by the glass of the telescope. And, as instruments have grown more complex, this issue has only grown more relevant. Researchers now routinely use high-tech machines such as electron microscopes, particle accelerators and DNA sequencing machines to make discoveries. But what these instruments reveal isn't unambiguous. Distinguishing between meaningless static and a meaningful signal can be very difficult. It requires interpretation, and people can disagree sharply about what the correct interpretation is.

Frank was convinced that the specks in the satellite images were a meaningful signal, and, to his credit, he diligently tried to gather more evidence to support his cause. For one brief moment, it even looked like he would be vindicated. In 1997, he triumphantly revealed that images from a second NASA satellite, Polar, showed identical dark specks in the atmosphere. Some of his staunchest critics admitted this gave them pause. It was definitely odd that these specks kept showing up. Perhaps they weren't just random instrument noise.

However, a subsequent analysis by an operator of the Polar camera revealed that the specks didn't change in size as the satellite's altitude varied. This suggested they were, in fact, an artefact. The operator speculated that they might be a result of static in the camera systems. As far as most scientists were concerned, this brought the controversy to an end.

Frank never gave up though. He kept on arguing for the existence of the small comets right up until his death in 2014. Without an advocate, the small comets then faded into obscurity. Unless one counts the specks themselves as a kind of quiet advocate, because they're still there, refusing to go away, waiting for someone to believe in them and transform them, Pinocchio-like, into real-life comets.

What if the Earth is expanding?

According to astronomers, the universe is expanding. In fact, its rate of expansion is accelerating as the force of a mysterious dark energy flings galaxies apart at an ever-increasing speed. The sun, they say, is also steadily growing larger as it burns through its hydrogen fuel, causing its core to grow hotter and its outer layers to swell. In about six billion years, the sun will be so big that it will entirely engulf Mercury and Venus, rendering the Earth uninhabitable.

Given this cosmic trend of expansion, could the Earth also be growing larger? At first blush, that might seem unlikely, since it's a rocky planet and rocks typically don't change in size. But there is a hypothesis which has hovered on the outer margins of geology for the past century that maintains this is exactly the case: the Earth is expanding and will continue to do so into the foreseeable future.

The hypothesis originated in the early twentieth century with the observation that the continents seem to fit together like pieces of a jigsaw puzzle. In particular, the coastline of South America matches that of West Africa almost perfectly.

Around 1910, the German meteorologist Alfred Wegener had become intrigued by this fit, and he came up with an explanation for it, arguing that, millions of years ago, the continents must have been clustered together as a supercontinent before subsequently drifting apart into their present locations. He called this his theory of continental drift. It received a frosty reception from geologists, however, partly because it contradicted the prevailing belief that the continents were permanently fixed in place, and partly because Wegener couldn't offer a convincing explanation for how the continents could possibly move through the dense material of the ocean floors. So, his theory was relegated to the radical fringe of geology.*

But Wegener wasn't the only one intrigued by the apparent fit of the continents. Even as his theory was suffering the ignominy of rejection, denounced as pseudoscience by leading geologists, a ragtag group of iconoclasts – including the German geophysicist Ott Hilgenberg in the 1930s and the Australian geologist S. Warren Carey in the 1950s – came to believe he had identified an important problem. Although they didn't think his solution of drifting continents was a very good one. Instead, they produced an entirely different explanation: the land masses might once have been joined if the entire Earth had originally been much smaller. And so was born the expanding-Earth hypothesis.

The basic concept of it, as envisioned by Hilgenberg and Carey, was that when the Earth first formed, it had been about half its current size of approximately 7,900 miles in diameter, and that, as the hot molten planet cooled, an unbroken rocky crust formed around its surface. In this way, the land masses of the Earth, in their original, primordial state, entirely enclosed the surface of the planet.

Then the planet began to expand. Why? We'll get to that soon, but for now just imagine the pressure of this expansion fracturing the rocky crust into pieces, creating the continents. These then spread apart as the Earth continued to swell. Magma welled up in the cracks created by the stress of this expansion, and this formed the ocean floor in the ever-widening gaps between the continents.

So, while Wegener had the continents moving across the surface of a fixed-dimension globe, like giant battleships ploughing

* See 'Weird became true: continental drift' for more about Wegener's theory.

through the dense material of the ocean floor, the expansion model didn't have them moving through anything at all. They remained in the same place and only moved vertically upwards, while the creation of new ocean floors enlarged the distances between them. As such, it was a model of continental spread, rather than drift.

One piece of evidence above all else had convinced Hilgenberg and Carey that their Earth-expansion hypothesis was superior to continental drift. They ardently believed that the continents fitted together much better on a smaller Earth.

It's a simple enough experiment to try: make the continents into jigsaw pieces, but do it on a globe, and then try to piece them together. You'll discover that, despite the overall appearance of matching contours, particularly the matching coastlines on either side of the Atlantic Ocean, they don't fit together perfectly. Large, triangular gaps yawn between them. Geologists refer to these gaps as 'gores', which gives an appropriately villainous feeling to them. Wegener's critics were fond of pointing out this poor fit in an attempt to debunk his theory.

What Hilgenberg and Carey discovered was that, if you keep your jigsaw continent pieces the same size, but try to fit them together on a smaller globe, it works better. The gores gradually disappear. If you make the globe small enough, the continents will almost seamlessly enclose the entire surface of the simulated planet. Hilgenberg in particular was an enthusiastic maker of globes of different sizes – his preferred medium being papier mâché – and globe-making became the signature craft tradition of the expansionist movement. Whenever its devotees met up at conferences around the world, they would proudly display their multi-sized globes to each other.

The expanding-Earth advocates did provide other reasons in support of the hypothesis. They pointed out that, unlike any other geological model at the time, theirs explained why there were two distinct types of surface on the Earth: the land and the ocean floor. They also noted that the expanding-Earth model kept the continents fairly evenly distributed across the globe, unlike Wegener's model, which created what they felt was a crazily lopsided Earth, with a giant supercontinent massed on one side of the planet and nothing but empty ocean on the other. But it was really the issue of the better fit of the continents on a smaller globe that most inspired them. It seemed utterly impossible in their mind that the contours of the continents would match so well in this circumstance just by coincidence.

But what in the world (literally) could have caused the Earth to expand? We've now arrived at this enigma. It was an obvious question that couldn't be avoided. Advocates knew that the mechanism of expansion was potentially the great Achilles heel of their hypothesis, so they expended enormous mental energy trying to dream up explanations for it.

The most scientifically conservative of these (which is to say, the one that least outraged scientists) was developed by the Hungarian geophysicist László Egyed in the 1950s. He proposed that material at the border between the Earth's core and mantle might be expanding as it underwent a phase transition from a high- to a low-density state, somewhat analogous to the way that water expands in volume as it undergoes the phase transition from liquid to solid ice. If this were occurring around the core, it would push upwards on the mantle, causing a very slow, gradual expansion of the Earth, at the rate of about one millimetre a year.

Other expansionists weren't content with such a slow growth rate, and they sought out mechanisms that involved far more speculative physics. Carey suggested that new matter might be coming into existence within the Earth's core, thereby increasing the mass and volume of the planet. He lifted this idea from the steady-state theory of cosmology, which envisioned the continuous creation of matter as the cause of the expansion of the entire universe.* At the time (the 1950s), the steady state was still

* See 'What if the Big Bang never happened?'

considered to be a serious rival to the Big Bang theory and had a number of prominent supporters. So, it seemed reasonable to Carey to draw upon it to explain the expanding Earth. Doing so also suggested a parallel between the expansion of the universe and the Earth.

An even more radical proposal was that the force of gravity might be declining throughout the entire universe. If this were so, it would cause the Earth's mantle to weigh progressively less, which would reduce the compressive forces on the core and thereby allow it to expand, like a spring returning to its previous size. This idea was the brainchild of the German physicist Pascual Jordan, who in turn had adapted it from the speculative musings of the British physicist Paul Dirac, who had suggested that the physical constants of the universe (all those numbers that are never observed to change, such as the speed of light, the mass of a proton and the force of gravity relative to mass and distance) might vary over time. Since physicists don't know why any of these constants have the values they do, it seemed vaguely plausible that some of them, such as the force of gravity, might change as the cosmos aged.

But a significant faction within the expansionist movement acknowledged that there really wasn't any obvious mechanism of growth, and they argued that this shouldn't be held against the hypothesis. After all, in science, the observation that something is happening often precedes the knowledge of why it's happening. Darwin described the process of evolution by natural selection long before biologists uncovered the genetic mechanism that allows it to occur. Perhaps, in the future, physicists would discover a mechanism that caused Earth expansion, but for now it was enough to note that it was happening, even if no one knew why.

This was the situation in the late 1950s. Continental drift and Earth expansion both offered rival explanations of why the continents seemed to fit together, but neither one seemed inherently more plausible than the other. Both demanded the acceptance of some highly counter-intuitive notions, and, as far as most geologists were concerned, both were equally absurd because the continents simply didn't move at all.

Then geological dogma was entirely upended. The cause of this dramatic turn of events was the discovery of sea-floor spreading. Sonar mapping of the ocean floor during the 1950s and '60s gradually began to reveal the existence of gigantic underwater rift valleys that ran throughout the oceans of the world. In these valleys, the sea floor was literally splitting apart and spreading outwards on either side, causing the formation of new oceanic crust where magma bubbled up in the ever-widening gap.

This was an astounding discovery. The rift valleys were massive as well as being highly dynamic geological features. It took only a few years for geologists to realize that the conventional wisdom of fixed continents had been shattered. The land masses obviously had to move. But the question was, which explanation of continental movement was most compatible with this new discovery: drift or expansion?

A number of leading scientists, such as Bruce Heezen, who had conducted much of the sea-floor mapping, thought it was expansion. After all, the expanding-Earth hypothesis had predicted the existence of tension cracks in the ocean floor where new oceanic crust would be produced. So, for a brief historical moment, it seemed as if expansionism had a real chance of achieving mainstream acceptance.

But it didn't last for long. Geologists soon realized that, not only was new crust being created in the rift valleys, but old crust was simultaneously being destroyed at subduction zones, located along the edges of the continents where the ocean floor plunged beneath the land. By 1970, this had led to the development of plate tectonics, which envisioned the creation and destruction of the ocean floor acting as a kind of conveyor belt that moved the continents endlessly around the globe. In this model, the continents didn't drift so much as float on top of the convection currents within the mantle.

This wasn't exactly like continental drift, but it was close enough to essentially vindicate Wegener. It was a derivation of his theory that emerged out of the discovery of sea-floor spreading as the new geological orthodoxy. The expanding Earth, on the other hand, still had no mechanism. Nor could its advocates explain the existence of subduction. So, after its brief flirtation with mainstream acceptance, its scientific status plummeted down like a rock.

That might seem like it should have been the end of the story of the expanding-Earth hypothesis. In hindsight, its advocates had correctly identified the fit of the continents as an important phenomenon, they'd just chosen the wrong explanation for it. Wegener had won the debate, not them. So, surely, the hypothesis would quickly fade away. Geologists certainly hoped that would happen.

But it didn't. Instead, the expanding-Earth hypothesis entered a new, altogether stranger phase of its career. Far from disappearing, it stubbornly hung around to become the great contrarian hypothesis of geology. Its advocates continued to meet up at conventions, they submitted articles to academic journals and, with the emergence of the Internet in the 1990s, they held court online to an apparently large and appreciative audience. All of which frustrated mainstream geologists no end.

It's worth noting that its supporters weren't an entirely homogeneous group; there was a spectrum of belief among them. On one end were the fast expansionists who argued that the Earth was growing at a rapid rate, well over five millimetres a year, and that most of this growth had occurred in the past 200 million years. The leading proponent of this view was the geologist James Maxlow, a student of S. Warren Carey, who predicted that, in another 500 million years, the Earth will have swelled into a gas giant the size of Jupiter.

At the other end of the range were the slow expansionists, who harkened back to the work of László Egyed. They argued that the Earth had been growing almost imperceptibly over its entire history, by mere fractions of a millimetre every year. The fast expansionists attracted the majority of the public attention because of their more extravagant claims, but, as far as most geologists were concerned, the whole bunch of them were equally crazy because there was simply no credible explanation for why the Earth would expand.

Given this, why did support for the hypothesis endure? One of the core issues motivating its advocates was their continuing conviction that the continents fitted better on a smaller globe. They returned to this subject obsessively, insisting that this superior fit shouldn't be possible by mere chance.

The most rigorous advancement of this argument was made by Hugh Owen, a palaeontologist at the British Museum. In 1984, Cambridge University Press published his *Atlas of Continental Displacement*. This consisted of an exhaustive cartographic tracking of what he maintained was the progressively better fit of the continents on a smaller Earth. The best fit of all, he concluded, was on an Earth 80 per cent its current size. Even mainstream reviewers, such as the British geologist Anthony Hallam, grudgingly acknowledged it was a work of serious, if eccentric, scholarship.

Perhaps the primary reason for the persistence of the hypothesis, however, was because all the evidence for or against it was somewhat indirect. The debate circled endlessly around questions such as the fit of the continents or possible mechanisms of expansion. What it didn't address was whether the Earth was actually measurably changing in size. This was because scientists couldn't put a tape measure around the planet to find this out, and the inability to do this created just enough doubt to allow the expansionists to continue to press their case.

At least, scientists couldn't do this until the twenty-first century, when satellite-based technology did actually make it possible. The expansion of the Earth could be put to direct empirical test, and a team of NASA researchers set out to do this.

Even with the advanced technology, it was a challenging task given that the Earth isn't a static entity. Sea levels and land masses are constantly in motion, rising and falling due to mountain formation and other natural processes, all of which had to be taken into account. But, in 2011, after a ten-year period of observation, the team reported their results: there was no evidence that the Earth was expanding. It appeared to be fixed at its current size. NASA issued a press release to spread the news, triumphantly declaring, in so many words, that the expansionist heresy could finally be put to rest.

And that, one might think, was that. Once again, it seemed the story of the expanding Earth had reached an end.

Or maybe not. Its supporters have proven to be quite persistent.

In 2016, expansion advocate Matthew Edwards published an article in the peer-reviewed journal *History of Geo- and Space Sciences* in which he acknowledged that the satellite measurements were indeed a problem for proponents of fast expansion. Rapid growth should definitely have been detected. But slow expansion of the type imagined by László Egyed, he argued, was a different story, particularly if one looked more closely at the results of the NASA study.

As it turned out, the researchers hadn't actually found no expansion at all. They had recorded expansion of 0.1 millimetres a year, possibly as high as 0.2 millimetres due to the inherent uncertainty of the measurement techniques. They deemed this to be 'not statistically different from zero'. But Edwards countered that it was absolutely different from zero, because even an expansion of 0.1 or 0.2 millimetres a year, over 4.5 billion years, becomes quite significant. And more recent measurements from Chinese researchers at Wuhan University had suggested that the expansion rate might be nudged up as high as 0.4 millimetres a year, which put it well within the range predicted by slow expansionists.

Had the expansionists snatched victory from the jaws of defeat? Edwards conceded that more study is required. One problem is that the vast majority of ground-based stations the satellites relied on to make measurements were located in the northern hemisphere. Perhaps future research, looking more evenly at the surface of the entire planet, will conclude definitively that no growth is occurring. But, even so, he insisted, it might be premature to close the case entirely on Earth expansion. A variety of it could yet be vindicated.

CHAPTER THREE

It's Alive!

As we approached the solar system, we noticed that the Earth differed dramatically in appearance from its planetary neighbours, hanging in space like a brilliant blue-green marble. The liquid water that covers much of its surface is the primary cause of this distinctiveness, responsible for the blue colour, but it's the presence of life that adds the tantalizing hint of green, and it is life which will be our focus in this section.

But what do we actually mean by 'life'? Scientists have struggled to answer this seemingly simple question. They've identified various properties that living things seem to uniquely possess, such as that they're highly organized, grow, reproduce, metabolize energy, respond to stimuli and evolve. But not everything we would identify as living possesses all these properties. Sterile animals can't reproduce, but they're nevertheless alive. And, more problematically, things we would assume to be non-living share many of these traits as well. Both crystals and fire, for example, can grow and reproduce. As a result of these ambiguities, researchers haven't been able to agree on a single definition of life. Many argue that it's a mistake to even assume there is a clear-cut division between life and non-life.

Returning to the view of our planet, other mysteries about life present themselves, such as how it began and why the Earth appears to be its only home. Trying to solve these puzzles by looking back in time to when the Earth first formed only raises more questions. Scientists date the formation of the Earth to approximately 4.54 billion years ago, and, relatively soon after that, around 3.7 billion years ago, we can see that living organisms were present in the primeval oceans. That's the enigma. Based on the evidence left behind, it's not at all clear how the life forms got there.

You might assume that it would primarily be biologists who ponder these issues of life's nature and origin. But, actually, interest in them is highly interdisciplinary, attracting the attention of cosmologists, astronomers, philosophers, physicists, geologists and even sanitation engineers. And there is very little scientific consensus about the answers. Some speculations are considered to be more plausible than others, but there are many that roam much further afield.

What if everything is conscious?

How can a bunch of inanimate matter transform into a living organism? That's the big question posed by the origin of life. Inanimate matter just sits there until external forces prompt it to move, whereas living organisms, even the simplest of them, possess agency. They do things. They metabolize nutrients, respond to stimuli, grow, reproduce and evolve. The gulf between life and non-life seems almost unbridgeable. But what if the central premise of this mystery, as framed by science, isn't quite right, because what if matter, in its fundamental state, isn't entirely inanimate? What if it possesses some qualities usually attributed only to living things? In particular, what if it possesses a rudimentary consciousness – an awareness of some kind that it exists?

This is the odd claim of the theory of panpsychism. Its advocates argue that absolutely everything in the universe is conscious. This includes people, dogs, plants, rocks, plastic bags, car tyres, iron nuggets, puddles of water and even electrons. To be clear, the theory doesn't claim that everything is conscious in the same way. The ability of an electron to experience or feel would obviously be very different, and far more primitive, than that of a human. Nor does the theory claim that everything has the power to reason. It doesn't imagine that rocks and car tyres are sitting around contemplating philosophy. Nevertheless, panpsychism does insist that all matter, at some level, has the capacity to experience its existence.

The name of the theory comes from the Greek words *pan* (all) and *psyche* (mind), and, while the concept may sound suspiciously like New Age mumbo jumbo, it's actually been around in Western culture for millennia and has, in the past century, attracted some prestigious supporters, including the physicist Sir Arthur Eddington and the philosopher Bertrand Russell. Most recently, the philosophers David Chalmers and Philip Goff have championed it.

When people first hear or read about panpsychism, their initial reaction is typically to dismiss it as absurd. This is the hurdle the theory always has to overcome, because it just seems so blindingly obvious that *everything* can't be conscious. Is anyone supposed to seriously believe that spoons and floor tiles are sentient? The possibility scarcely seems worth debating. It's self-evidently ridiculous.

But defenders of the theory point out that, paradoxical as it might seem, for most of history, going all the way back to pre-historic times, panpsychism has been widely regarded as the common-sense point of view to take. The assumption that non-biological matter lacks consciousness actually represents a new, peculiarly modern way of looking at the world.

For example, animism is a form of panpsychism. This is the belief that every aspect of the natural world – the sun, wind, water, rocks and trees – is animated by its own spirit. Anthropologists suspect that this was a universal feature of the earliest forms of religion practised by hunter-gatherer societies.

Animist assumptions continued to appear in the thoughts of the earliest Greek philosophers, such as Thales of Miletus and Pythagoras, who all took panpsychism for granted. In the fourth century BC, these ideas became enshrined in the writings of Aristotle, whose work served as the basis for the Western understanding of the natural world for the next 2,000 years.

In his work, Aristotle explained the behaviour of non-living things by assuming that they exhibited some of the qualities of living things. Specifically, a sense of purpose and therefore a primitive form of consciousness. According to him, everything had a proper place or 'final cause' that it strived to achieve. For example, fire and air floated upwards, he said, because these elements naturally belonged in the sky, and so they strived to attain their correct location. An inner will propelled them. Likewise, earth and water belonged at the centre of the world, and so they strived to move downward.

Historians of science note that, in this way, Aristotle's universe was fundamentally biological. He used the properties of life, such as purpose and will, to explain the behaviour of non-life. Everything in Aristotle's universe, including astronomy and chemistry, ultimately reduced to biology. But, around the seventeenth century, a new philosophical world view called mechanism emerged, which insisted that everything, in its fundamental state, consisted of inert matter entirely lacking a sense of inner purpose. Everything, in this way of thinking, ultimately reduced to physics. Even biological organisms were reimagined as being like complicated machines.

Why this new world view emerged at that particular time, in the particular place it did (Europe), is a matter of debate. One theory is that it might have had something to do with the invention of weight-driven mechanical clocks. By the middle of the fourteenth century, such clocks had been installed in the central squares of most major cities. On the outside they appeared animated, but everyone knew that on the inside they were just a bunch of inanimate metal cogs and gears. So, they offered a kind of ready-made metaphor for how nature might operate. They demonstrated how lifeless matter, arranged in the right way, might give the appearance of being alive. By the seventeenth century, it had become popular to compare nature to a clock and to describe God as a watchmaker.

Whatever may have been the reason for the emergence of this new mechanistic world view – and it doubtless involved more than just clocks – it turned the premise of Aristotle's ancient science on its head. The universe was now assumed to be made up of 'dead' particles, entirely lacking any kind of animating spirit and set in motion by mechanical forces (like the weights that made the gears of a clock move). As the eighteenth-century philosopher Immanuel Kant put it, 'lifelessness, *inertia*, constitutes the essential character of matter.' This was the great change in perspective that ushered in the rise of modern science.

This world view framed the origin of life as a particular problem, because it now had to be explained in terms of non-life, rather than vice versa. Researchers had to explain how what was entirely inanimate could somehow become animate.

But, although panpsychism may have been eclipsed by mechanism, it evidently still lingers in the cultural memory. Philosophers have noted that the concept of 'the Force', featured in the *Star Wars* movies, described as a living energy that connects everything in the universe, is essentially panpsychism. In the 2011 UK census, 176,632 people claimed to be practitioners of the Jedi religion, and so would be believers in the Force. Jediism, in fact, ranked as the seventh most widely practised religion in the UK.

That's all well and good, one might say. Perhaps panpsychism represents an old way of understanding nature, but people didn't know any better, back then, and we've now corrected the naive assumptions of our ancestors – not counting those Jedi among us (whose professed belief may, anyway, be more tongue-in-cheek than genuine). The proof of the accuracy of our present-day view lies in all that modern science has accomplished. We've put men on the moon, built supercomputers and unlocked the genetic code.

Proponents of panpsychism concede that the mechanistic world view was an important development in Western thought, but they argue that it has now, in turn, become limiting, because it's bumping up against phenomena it can't explain. In particular, it struggles to explain consciousness.

The problem, they say, is that physics only explains the behaviour of things. It reduces the natural world to a series of forces – such as electromagnetism and gravity – that act on particles, modifying how they behave in various ways. These behaviours are all external properties and are susceptible to measurement. What physics doesn't examine, or even attempt to explain, is the inner nature of particles, what they intrinsically are.

Take yourself, for example. If physicists were to describe you, they might point out the complex forces acting on your body and how they cause you to behave. There's the gravity pulling down on you, the electrical signals running along your nerves, and the chemical reactions within your cells. But you're more than just these external measurable properties. You also have an inner life, or consciousness. You see colours, feel pain and smell aromas. You experience things such as fear, anger and happiness. You have a subjective awareness of reality. All this is invisible to physics. Left to its own devices, physics would never be able to predict that a phenomenon such as consciousness even existed.

So, the panpsychic argument goes, if physics can't fully describe a person, because it overlooks their inner being, then why do we assume that physics would be able to describe fully any other form of matter? Or, to put it another way, why should we believe that matter can be fully described by measuring only its external properties, as scientists claim, if we know this isn't true for ourselves? Instead, it would make more sense to assume that *all* matter has some kind of interior perspective, just as we do – that it has some kind of awareness of its surroundings. In which case, it must feel like something to be a rock or an electron, even though that feeling would be completely alien to our experience.

The scientific response to this argument is that we can be pretty sure electrons, rocks and tables aren't conscious because they display no mind-like behaviour. All the evidence indicates that consciousness is a special feature of brains, and brains alone.

Furthermore, they insist, there's a perfectly good explanation of consciousness, grounded in physics. Atoms form into molecules, which form into neurons, which then form into brains. Consciousness emerges when enough neurons are collected together in a brain and start interacting with each other. The details of this are admittedly murky. Researchers aren't sure how many neurons are necessary, or what exactly about their arrangement triggers consciousness, but the overall picture seems clear enough. Consciousness is an emergent property of neuron communication. Or, put more plainly, it's a result of clever wiring.

Panpsychists, however, have another line of attack. They argue that while this explanation of consciousness may sound reasonable in principle, it actually poses a logical paradox because it suggests that non-aware particles can, merely by arranging themselves in a certain way, achieve awareness. But how can awareness emerge where there was no awareness before? If that were true, it would imply that something was emerging from nothing, which is impossible.

By way of analogy, consider electricity. It assumes many forms throughout the universe. It flows through wires in your walls as current, flashes through the sky as lightning and makes your hair stand on end as static electricity. These different electrical phenomena emerge as atoms arrange in various ways. However, they don't emerge out of nothing. They're a consequence of the fact that, at the most fundamental level, subatomic particles such as electrons have the property of either positive or negative charge.

Or consider gravity. It causes the emergence of many different phenomena, such as planets, suns, galaxies and black holes. But, again, these phenomena don't emerge out of nothing. Gravity is a property of particles at the smallest level of matter.

But when it comes to consciousness, unlike electricity and gravity, there is (according to science) no specific subatomic property out of which it emerges. It simply springs into existence, fully formed, as a consequence of atoms being arranged into brains. Where there was no awareness before, suddenly it's there. Panpsychists protest that this makes no sense. It would be like claiming that lightning could emerge into existence even if electrons didn't possess the property of charge. Or that planets could form even if gravity wasn't a universal property of matter.

Therefore, panpsychists argue that some form of primitive consciousness must be a property of subatomic particles. If this were the case, it would eliminate the paradox of awareness seeming to emerge into existence out of non-awareness. The consciousness that we as humans experience could then be seen as merely one manifestation of the consciousness inherent in all matter, in the same way that lightning is one manifestation of the force of electromagnetism inherent in all matter.

That's the basic argument for panpsychism, but critics offer a battery of reasons why no one should believe it. The most frequently repeated reason focuses on the so-called 'combination problem'. If every individual subatomic particle possesses its own consciousness, ask critics, then how do all these tiny consciousnesses combine to form a larger unified consciousness, such as a brain? Because it seems to be a distinctive feature of consciousness that it doesn't combine. If you put twenty people together in a room, their brains don't meld together to form a single mega-brain. So, why would electron consciousnesses combine seamlessly together?

But the most fundamental problem of all, according to critics, is simply that panpsychism is useless. It's completely untestable, which makes it unscientific, and it also generates no predictions. The neuropsychologist Nicholas Humphrey has complained that the theory 'crumbles to nothing' when asked to do any kind of useful explanatory work. After all, scientists could argue until they're blue in the face about what the inner being of an electron is, but the fact is, for better or worse, they can only ever know its external properties. They gain nothing by attributing consciousness to it.

Panpsychists counter that the same argument could be used against the mechanistic view of matter. It's equally unprovable that matter is mindless and lifeless. Moreover, they contend, there actually are some things to be gained by adopting a panpsychic outlook. It arguably offers a more naturalistic way of understanding consciousness because it fully integrates the phenomenon into nature, whereas the mechanistic view assumes there's a radical difference between non-conscious matter and the conscious mind.

Shifting to a moralistic line of argument, they also maintain that panpsychism offers a more spiritually enriching perspective on the universe. They accuse the mechanistic world view of having de-animated nature by reducing it to mere mechanical interactions. This, they allege, has contributed to present-day social and environmental problems by facilitating the exploitation of natural resources. As David Skrbina, author of *Panpsychism in the West*, has put it, 'It is easy to abuse dead, inanimate matter, or unconscious forms of life.' So, panpsychism is presented as a more compassionate and sustainable philosophy.

But having said all that, at the end of the day, panpsychists do acknowledge that the mechanistic world view is so deeply ingrained in modern society that it might be difficult to convince many people to abandon it. The panpsychists themselves admit to having moments of doubt, moments when they wonder if it's worth it to have their colleagues think they're crazy. The Finnish philosopher Paavo Pylkkänen has even coined a medical term to describe such doubts. He calls it *panphobia*, which he defines as the fear a person feels when he realizes that he actually finds the arguments of panpsychism compelling. Alternatively, if you do find yourself drawn towards belief in a panpsychic universe, you might be cheered at the thought of becoming a Jedi. Although, unfortunately, that won't get you a lightsaber or powers of levitation.

What if diseases come from space?

When there's a cold going around, it always seems as if you end up catching it no matter what precautions you take. You may shun runny-nosed friends, scrupulously avoid touching doorknobs, clean your hands constantly with antiviral sanitizer and even wear a face mask when on public transport – but no luck. You get sick anyway.

What did you do wrong? Perhaps you didn't cast your net wide enough when considering the possible sources of infection. You probably assumed that you would catch the cold from other people, but what if the germs were coming from a more exotic source? What if they were raining down upon you from outer space?

In 1977, the astrophysicists Sir Fred Hoyle and Chandra Wickramasinghe proposed that exactly this was the case. This was the same Fred Hoyle whom we met earlier as the co-author of the steady-state theory. Now, at a later stage in his career – in his early sixties, when most people might be looking ahead to retirement – he had joined forces with Wickramasinghe, his former student at Cambridge, to argue that passing comets were shedding pathogens which then drifted down into the Earth's atmosphere and caused outbreaks of disease. And not just the common cold. The two theorists speculated that many of the great epidemics of history, such as the fifth-century-BC plague of Athens, the Black Death of the fourteenth century and the flu pandemic of 1918, might have been caused by these tiny extraterrestrial invaders.

If true, this alone would be of great significance for humankind, but Hoyle and Wickramasinghe imagined implications that extended far beyond the history of medicine. They maintained that pathogens have been raining down from outer space for billions of years, and that this was the reason life first emerged on Earth. In other words, they claimed that we are all the descendants of space germs.

What Hoyle and Wickramasinghe were proposing was a form of panspermia, which is the idea that life didn't originate on Earth, but instead arrived here from elsewhere in the cosmos.

The prevailing view among scientists is that life began right here on Earth – perhaps, as Charles Darwin put it in 1861, in some 'warm little pond with all sorts of ammonia and phosphoric salts, light, heat, electricity etcetera present.' This seems like a reasonable assumption given that the Earth remains the only place we know of in the universe where life exists. Nevertheless, panspermic notions have been kicking around for a long time in Western culture. In ancient Greece, the philosopher Anaxagoras suggested that life originally fell to Earth from the sky, and in the nineteenth century leading scientists such as Lord Kelvin and Hermann von Helmholtz advocated versions of this idea. Kelvin compared it to the way that seeds blown by the wind take root in the scorched earth surrounding a volcanic eruption. In 1906, the Swedish chemist and Nobel Prize-winner Svante Arrhenius gave the concept its name, adapting it from the Latin phrase *panspermia rerum*, meaning 'the universal seed of things'.

Panspermia theorists have imagined life arriving on Earth in a variety of ways. Arrhenius speculated that tiny bacterial spores might be able to float up into the upper atmosphere of a planet and from there drift into space, where the force of solar radiation would propel them throughout the galaxy until some arrived here on Earth. Others believe that microbes from other planets could have arrived here inside rocky meteors. In 1960, the physicist Thomas Gold even suggested that aliens might have landed a spaceship on our planet sometime in its early history and then accidentally left behind microbe-contaminated garbage from which all life on Earth then evolved.*

The notion of panspermia has struggled to achieve mainstream acceptance, though, because critics have complained that it doesn't actually address the problem of how life emerged. It simply kicks the problem down the road, removing it from Earth to somewhere else. Many scientists also believe that it would be very difficult, if not impossible, for organisms to survive for extended periods while travelling through the vacuum of space.

Hoyle and Wickramasinghe addressed both these concerns simultaneously. They argued not only that life could survive in deep space, but that it had originated there – not on the surface of a planet, but in the vast clouds of dust that span the distances between the stars.

Wickramasinghe had been studying interstellar dust, trying to figure out just what the stuff was, and it was from this research that the germs-from-space theory evolved. When he began his research in the 1960s, the prevailing assumption had been that the dust mostly consisted of dirty ice (frozen water mixed with a few metals), but, by analysing the wavelength of radiation that the dust absorbed, Wickramasinghe concluded that this wasn't correct. Instead, and quite surprisingly, its absorption spectrum seemed to be a good match for a range of organic compounds that included the plant matter cellulose.

Organic compounds are loosely defined as molecules containing carbon. They're the material out of which living things are made. In the scientific effort to discover the origin of life, great

---

* Yes, this is the same Thomas Gold who co-authored the steady-state theory with Hoyle. See 'What if the Big Bang never happened?' And we'll meet him again soon in another chapter!

emphasis has been placed on understanding ways in which organic compounds might have formed out of inorganic matter, because, the thinking goes, once you have organic compounds, you have the building blocks of life. All that remains is to figure out how these building blocks arranged themselves in the appropriate way.

In 1953, the chemist Harold Urey and his graduate student Stanley Miller made headlines by designing a simple experiment that appeared to show that the atmospheric conditions on the early Earth might have allowed organic compounds to form quite easily out of inorganic materials such as methane, hydrogen and ammonia. The implication of this research was that the early-Earth environment must have given birth to life. This was widely hailed as the first real scientific breakthrough in understanding the origin of life.

However, Wickramasinghe's research was indicating that organic compounds existed in vast quantities in deep space. He shared these results with Hoyle, which prompted him to realize that Urey and Miller's assumption that the building blocks of life had first formed on Earth was mistaken.

Hoyle suggested that, if organic compounds exist in interstellar dust, then this, rather than the Earth, might logically be the place where life first formed. This was the Eureka moment for the germs-from-space theory and the observation that got both researchers hooked on the idea. But there was an obvious problem, which they acknowledged. Interstellar dust is very cold. Temperatures out there in the dust clouds hover close to absolute zero. It's not the kind of environment, one imagines, that would generate the complex chemical interactions associated with life.

The two researchers concluded, therefore, that life must not have emerged directly in the dust itself, but rather in the comets that were created as a dust cloud collapsed to form our solar system. In the interior of these comets, they pointed out, there would have been all the necessary ingredients for life: organic compounds, water, protection from ultraviolet radiation, and possibly even heat created by the decay of radioactive elements. Plus, there were billions of these comets, a vast chemical laboratory in which nature could have tinkered until it found the formula for life.

Once life had formed in a single comet, they argued, that comet would have begun scattering cells as it orbited the sun, spreading organisms to other comets, which would then, in turn, have scattered them even more widely. In this way, life forms would have eventually drifted down upon the newly formed planets in the solar system, taking root wherever they could, and these life forms would have found a particularly welcoming home on the young Earth.

The two researchers became obsessed by their germs-from-space theory. Over the following decades, they churned out a series of books and articles on the subject, seeking to amass evidence to persuade the sceptics, a group that included most of the scientific community.

One of their arguments focused on the speed at which life had emerged on Earth. As far as scientists can tell, single-celled microorganisms appeared almost as soon as it was physically possible for this to happen, once the surface of the planet had cooled down sufficiently to allow it. This seems remarkable when one considers how complicated the chemistry of life is. Hundreds of thousands of molecules have to be arranged in exactly the right way for even the simplest cell to function. How did such staggering complexity manage to arise so quickly? Hoyle and Wickramasinghe countered that if you move the creation of life off the Earth, into space, that buys a lot more time for the process to have happened. It could potentially give life billions of years to have emerged, rather than just a few million.

They also pointed out that bacteria are amazingly hardy. Research has revealed that many can survive intense radiation. But how, the two authors asked, did bacteria ever evolve this ability, given they would never have been exposed to such high radiation levels on Earth? Again, this is a challenge for the Earth-origin theory to explain. Such resistance makes perfect sense, however, if bacteria first evolved in the high-radiation environment of outer space.

By the standards of mainstream science, these were perfectly legitimate points to raise, but, as Hoyle and Wickramasinghe thought through the implications of comets giving birth to life, they came up with other arguments that proved far more controversial. It occurred to them that the rain of organisms from comets wouldn't have stopped four billion years ago. It would have continued and could have had ongoing effects here on Earth. It could, for example, have interfered with the course of terrestrial evolution. They noted the presence of gaps in the fossil record, where evolution had seemingly taken sudden leaps forward. Perhaps these gaps indicated times when infusions of genetic material from space had caused abrupt changes in species. This suggestion outraged biologists, because it essentially called into question Darwin's theory of evolution by natural selection.

Then Hoyle and Wickramasinghe moved their argument forward to the present. After all, there was no obvious reason why the rain of comet organisms wouldn't still be happening today. This led them to suspect that outbreaks of infectious disease might be caused by extraterrestrial germs falling to Earth from comets, echoing ancient fears that these heavenly objects were harbingers of doom.

One somewhat whimsical piece of evidence they put forward in support of this claim involved the features of the nose. Perhaps nostrils had evolved to face down, they speculated, in order to prevent space germs from dropping into them!

On a more serious note, they pointed to a curious aspect of epidemic disease – how difficult it can sometimes be to pinpoint precisely where outbreaks begin, because they often seem to begin in multiple, geographically diverse locations simultaneously. For example, the first cases of the great flu pandemic of 1918 were reported simultaneously in India and Massachusetts. If germs travel from person to person, this is a puzzle. One would think the pattern of infection should show a steady radiation outward from a single point. Hoyle and Wickramasinghe argued that if, on the other hand, pathogens were floating down from outer space, these diverse points of origin were no mystery at all.

To bolster the case for this model of disease transmission, Hoyle and Wickramasinghe transformed themselves into amateur epidemiologists and conducted a study of an outbreak of flu that occurred in Welsh boarding schools in 1978. Based on their analysis, they concluded that the spread of the disease in the schools couldn't adequately be explained by person-to-person transmission, because the distribution of victims in the dormitories seemed entirely random. They argued that vertical transmission (germs falling from space) better explained the pattern of the outbreak.

They even suggested that it might be possible to connect outbreaks of disease to the passage of specific comets. Here, they drew particular attention to a rough periodicity in global outbreaks of whooping cough, which seem to occur, on average, every three and a half years. This correlated uncannily well with the regular return in the sky every 3.3 years of Encke's Comet.

By the late 1970s, Hoyle already had a reputation as a scientific troublemaker on account of his advocacy of the steady-state theory and consequent insistence that the Big Bang never happened. But, when he declared that the Earth was under constant attack from extraterrestrial pathogens, many wondered if he had completely taken leave of his senses. If he hadn't been such a prestigious astrophysicist, the scientific community would have simply ignored him and Wickramasinghe.

As it was, critics blasted just about every detail of the germs-from-space theory. Astronomers pointed out that comets are extremely hostile environments, so it defied belief to imagine life arising inside of them. The one thing that biologists believe to be absolutely essential for life is liquid water. All life on Earth depends on it, but it's doubtful that it could exist in liquid form within a comet. It would be frozen solid. And even if one assumes that heat from radioactive elements could maintain a liquid core inside a comet, the radiation itself would cook any organism. Either way, life wouldn't survive.

The astronomical critiques, however, were mild compared to the scorn and opprobrium that biologists poured down upon the theory. Sir Peter Medawar, winner of the 1960 Nobel Prize in physiology or medicine, denounced it as 'the silliest and most unconvincing quasi-scientific speculation yet put before the public.'

The aspect of the theory that particularly incensed biologists was the fact it didn't differentiate between bacterial and viral infection. The idea of bacteria forming inside comets seemed far-fetched, but it was at least vaguely within the realm of possibility. But *viruses* in comets? The notion seemed completely absurd, because viruses require a host to replicate and this wouldn't exist in a comet. It would be impossible for viruses ever to have evolved in such an environment.

There's actually an extraordinary specificity between a virus and its host. Each virus is highly evolved to attack only certain types of cells within certain species. How could such specificity ever have been acquired inside comets travelling through space, millions of miles away from the potential host cells on Earth?

Experts also dismissed Hoyle and Wickramasinghe's efforts at epidemiology as laughable, criticizing them for not taking into account factors such as varying degrees of pre-existing immunity or differences in the amount of pathogens shed by individuals, which can explain much of the unpredictability of patterns of disease transmission.

Hoyle and Wickramasinghe didn't help themselves when they started adding increasingly fantastical elements to their theory. They speculated that not only was interstellar dust made of organic compounds, but that it might actually consist of vast clouds of freeze-dried bacteria. They suggested that life was so improbable that its evolution must be directed by a cosmic intelligence that was somehow coordinating what types of genetic material to shower upon the Earth. And perhaps, they mused, it wasn't just bacteria and viruses falling from comets. In their 1981 book *Evolution from Space*, they argued that comets might also be dropping insect larvae into our atmosphere.

At which point, the scientific community ceased to take them seriously.

This may make it sound like the germs-from-space theory doesn't have a scientific leg to stand on. In terms of its reputation, this is probably true. Mention it to scientists and they typically react by rolling their eyes. But this isn't true about all the individual elements of the theory. Some parts have proven more scientifically robust.

For instance, Wickramasinghe's observation that interstellar dust contains vast amounts of organic compounds has been confirmed, and he gets credit for first making this discovery. At the time of writing, over 140 of these organic compounds have been identified. Most astronomers wouldn't agree that these compounds are specifically cellulose, and they definitely reject the claim that space dust is made up of freeze-dried bacteria. Nevertheless, the organic compounds are definitely out there, and scientists other than Hoyle and Wickramasinghe have concluded that this could potentially be very relevant to how life originated.

Many origin-of-life researchers now believe that these compounds might have kick-started the evolution of life by being deposited on the surface of the young Earth by comets. This scenario has been described as 'soft panspermia' because, while it doesn't involve life itself coming from space, it does imagine that the building blocks of life arrived from space rather than being fashioned on Earth, as Urey and Miller believed.

Even the idea that life could originally have formed inside a comet has received some experimental support. In the late 1990s, the astrochemist Louis Allamandola simulated the ultra-cold environment of space dust inside a special chamber at NASA's Ames Research Center. By doing so, he showed that when ultraviolet radiation struck molecules on the surface of the dust, it caused them to form into complex organic compounds. Even more intriguingly, Allamandola reported the formation of vesicles with cell-like membranes. This was significant because while organic compounds are essential to life, some kind of membrane to keep those molecules separated from the external environment is also necessary. Space dust could potentially have produced both.

This led Allamandola to speculate that billions of years ago, these cell-like vesicles and organic compounds might both have been embedded within the ice of comets. As the comets passed around the sun, they could have been warmed and jostled around just enough to cause some of the compounds to work their way into the vesicles, thereby producing the first incarnation of a living cell.

Allamandola acknowledged this was possibly a 'crazy idea'. He might have had the memory of the germs-from-space theory in the back of his mind. Regardless, his experimental data demonstrated that there is still a viable case to be made for the idea that life originated in comets. It hasn't been completely ruled out of contention. And, if life did originate in a comet and arrived on Earth by that means, could life forms still, on occasion, make their way to Earth from a comet? The idea isn't impossible. Perhaps there really are comet creatures in the family tree of life.

Weird became plausible: the vent hypothesis

In February 1977, marine geologist Jack Corliss of Oregon State University and two crewmates were cruising a mile and a half beneath the surface of the ocean, near the Galapagos Islands. They were in the *Alvin* submersible, a craft specially designed to withstand extreme depths, searching for a hydrothermal vent that had been remotely detected the previous year, which they wanted a closer look at. These vents are giant fissures in the ocean floor. Submarine geysers of scalding hot water, heated by contact with underground magma, blast out of them and mix with the freezing-cold water of the ocean.

They found the vent soon enough, but they also found something far stranger. Clustered around it was a thriving ecosystem that included massive clams, giant tube worms and even albino crabs, all living in complete darkness. Life was the last thing they had expected to find because, after all, they were 8,000 feet below the surface of the ocean. The pressure at a mere 2,500 feet is enough to crush a nuclear-class submarine. Biologists had assumed that the ocean floor at such depths would be sterile. In fact, the conventional wisdom at the time was that life could only survive within a relatively narrow temperature and pressure range, which the vent system was way outside of.

This discovery of the vent ecosystem is now regarded as one of the great moments in origin-of-life studies, because it led many to suspect that hydrothermal vents may have been where life began. But, when Corliss first proposed this idea following his return home, it was considered so dubious that most journals refused to publish it, and it took over a decade to gain acceptance.

Corliss didn't develop the idea on his own. It emerged out of discussions with several of his Oregon State colleagues (John Baross and graduate student Sarah Hoffman). What led them to the concept was that, in addition to being rich in energy, vents boast a number of potentially life-friendly features. For example, the temperature difference between the burning-hot geysers and the chilly surrounding water can facilitate chemical reactions of the kind that could have promoted the complex chemistry required for the emergence of life. Plus, such systems contain significant concentrations of methane, ammonia and hydrogen, from which organic prebiotic molecules, i.e. 'the building blocks of life', could have formed.

In 1979, the three researchers co-authored an article detailing this vent hypothesis. They thought it was an important contribution to the debate about the origin of life, but they soon discovered that their scientific peers didn't seem to agree. The leading scientific journals, *Nature* and *Science*, both promptly rejected it. Other editors similarly treated the hypothesis as too far-fetched to merit serious consideration.

The problem was that the vent hypothesis flatly contradicted the then-dominant 'prebiotic soup' model, which imagined a far gentler beginning for life. In this theory, organic molecules had first formed in the hydrogen-rich atmosphere of the early Earth and had then rained down into the oceans, where they mixed together in the warm waters like ingredients in a slowly simmering soup, before eventually combining to make living cells.

The Russian biochemist Alexander Oparin and the British biologist J. B. S. Haldane had both independently proposed versions of this theory in the 1920s. But it was in the early 1950s that the theory had really taken hold, when the Nobel Prize-winning chemist Harold Urey and his graduate student Stanley Miller conducted an experiment in which they shot a spark (meant to simulate lightning) through a mixture of what they assumed to have been the primary components of the Earth's early atmosphere: methane, hydrogen and ammonia gases. They found that, sure enough, organic molecules such as amino acids quickly formed.

The results of the Miller–Urey experiment seemed to offer dramatic confirmation of the prebiotic-soup model, and, based on the success of the experiment, Miller rose to become the undisputed dean of origin-of-life studies in the decades that followed. He trained students who then got jobs at leading universities, carrying with them the gospel of the model and further extending Miller's influence.

So, when the vent-hypothesis article was sent around by journals for peer review, it ran headlong into this academic wall of the prebiotic-soup faithful, to whom the idea that life might have arisen in an environment as violent and turbulent as a submarine hot spring seemed patently absurd. Miller and his disciples shot holes in the article, arguing that the extreme temperature of the vents would have caused organic molecules to quickly decompose. They also pointed out that the geochemistry of modern-day vent systems depends upon oxygen, which wouldn't have been present when life first formed, because oxygen was only produced later by photosynthesizing surface organisms. It also didn't help that Miller was known to be extremely protective of his status as the reigning master of origin-of-life studies.

The article did get published eventually, in 1981, in the obscure journal *Oceanologica Acta*. And it might have languished there forever, but, against all odds, it caught the attention of a few researchers and began to circulate around, gaining fans. This was in the days before the Internet. So, as the geologist Robert Hazen later recounted, photocopies of the article were passed from one researcher to another, defying the efforts of Miller and his disciples to block it.

The reason for the wider interest was that a few geologists had begun to have doubts about the prebiotic-soup model. Their research indicated that the Earth's primordial atmosphere hadn't consisted of methane, hydrogen and ammonia, but instead had mostly been gases belched out by volcanoes: nitrogen, carbon dioxide and water vapour. Organic molecules wouldn't have easily formed in such a climate, and this significantly weakened the case for a prebiotic soup. To the geologists, the vent hypothesis seemed like a plausible alternative.

When Miller and his disciples realized the vent hypothesis was attracting attention, they aggressively pushed back against it, mocking it as a fad and continuing to stress that the vents were simply too hot to have allowed life to originate. Miller even told a reporter from *Discover* magazine, 'The vent hypothesis is a real loser. I don't understand why we even have to discuss it.'

However, the Millerites couldn't stop the trickle of discoveries that kept adding weight to the opposing hypothesis, such as the realization by marine geologists that the vent found by Jack Corliss wasn't a one-off. Similar hydrothermal vent systems lined undersea ridges throughout the Atlantic and Pacific. They represented a vast, previously unknown environment that would have existed on the early Earth.

Supporters of the vent hypothesis also pointed out that the extreme depth of the vents could actually have been a benefit for early life, not a drawback, because any incipient life form would have been protected from surface hazards such as asteroid strikes and ultraviolet radiation.

But the crucial development that turned the tide in their favour occurred when biologists, inspired by the discovery of the vent ecosystems, started looking elsewhere for 'extremophiles', or organisms that thrive in extreme environments. They ended up finding them everywhere, surviving in the most unlikely conditions imaginable: in the frozen ices of Antarctica, in the bone-dry Atacama Desert of South America and even in rocks up to seven kilometres beneath the ground.

Among the most remarkable extremophiles are so-called tardigrades, aka 'water bears'. These are microscopic eight-legged creatures that live in almost every environment on Earth, from the deep sea to the tops of mountains. Research has revealed that they can survive near-complete dehydration, temperatures up to 150 degrees Celsius (300 degrees Fahrenheit) and even the vacuum of space. They could easily live through a nuclear apocalypse that would wipe humans out.

The discovery of such hardy organisms upended the old assumption, which the prebiotic-soup model had taken for granted, that life is fragile. It's not. It finds ways to thrive in a broad range of temperatures and pressures, and this suggests it might have originated in an environment we would consider to be lethal. By the 1990s, the growing realization of this had led to the vent hypothesis being widely accepted as a viable origin-of-life scenario – much to the consternation of the Millerites.

Of course, being accepted as plausible isn't the same as being considered the answer to the puzzle of life's origin. Far from it. If anything, an answer to that question seems further away than ever. Since the 1980s, the number of possible locations that researchers are willing to consider has steadily expanded. Could arid deserts have birthed life? Or perhaps frozen polar fissures filled with briny water? What about hot volcanic geysers? Or maybe the first cells formed within the pores of rocks, miles underground. All these speculations now have their supporters. A significant minority even champions an extraterrestrial origin. But it was the vent hypothesis, inspired by giant tube worms and albino crabs living a mile and a half deep on the ocean floor, that first challenged the long-held assumption that life must have originated somewhere gentle.

What if the Earth contains an inexhaustible supply of oil and gas?

Oil and natural gas play a crucial role in powering the world's industrial economies, but where do these remarkable substances come from? Out of the ground, of course! But have you ever thought about how they got down there in the first place?

The prevailing wisdom is that they're of biological origin, formed millions of years ago from biomass such as plankton and algae that was trapped underground, either by sediment depositing on top or by shifting tectonic plates. The pressure of the rocks above then heated and compressed the biomass, transforming it into fuel. That's why it's called fossil fuel, because it's believed to derive from the fossilized remains of life forms that previously inhabited the world. As such, the supply is presumably limited, because there was only so much biomass created over the course of the Earth's history. If we keep using more oil and gas, eventually we'll run out.

However, in the 1970s, the astrophysicist Thomas Gold of Cornell University emerged as the champion of an alternative theory in which he argued that these fossil fuels aren't of biological origin at all, but rather are 'abiogenic' (non-biologically made). He maintained that natural gas (methane) was part of the original matter from which the Earth first formed, that vast reservoirs of it were then sealed within the deep crust and mantle, and that it has been slowly but constantly seeping upwards, towards the surface, ever since. As it does so, he said, the geological forces of heat and pressure transform much of it into oil.

This theory led him to make the extraordinary claim that we're not in any danger of running out of these fuels. Not even close. He insisted that the supply is planet sized, and therefore, for all practical purposes, limitless. It's so plentiful that we might as well consider oil and gas to be renewable resources, in the same way that solar energy is. We could continue consuming them at our current rate for millions of years and not risk exhausting the supply.

To this he later added one more sensational claim. This abundant upwelling methane doesn't just exist, he said – it was also responsible for the origin of life.

You might recall that we've met Gold before.* In the 1940s, he co-authored the steady-state theory, according to which, new matter is continuously being created throughout the universe. But that was just one of his many radical ideas. Over the course of his career, he was prolific in dreaming up weird theories. If all his theories had remained fringe and unaccepted, the scientific community would have written him off. But, to the contrary, quite a few of his seemingly eccentric concepts were eventually vindicated, earning him a reputation as someone whose ideas it was unwise to dismiss too readily. Most famously, in the late 1960s, when astronomers first detected mysterious, rapidly repeating radio pulses coming from deep space, Gold proposed that the pulses could have been produced by fast-spinning stars made entirely from neutrons. Most scientists initially thought this idea was somewhat far-fetched, but within a year it was widely accepted, and remains so to this day. These stars are now known as pulsars.

The idea that oil and gas are of non-biological origin wasn't actually original to Gold. The concept had been kicking around

* See 'What if the Big Bang never happened?' and 'What if diseases come from space?'

for quite a while, having been proposed as early as the 1870s by the Russian chemist Dmitry Mendeleyev, creator of the periodic table. Throughout much of the twentieth century, it remained popular among Russian scientists, but in Europe and America it never caught on. By the 1940s, most Western geologists firmly believed that fossil fuels are of biological origin. Various chemical clues led them to this conclusion, such as the presence in oil of biological molecules such as lipids (organic fatty acids), resembling those found in bacterial cell membranes. They also primarily found oil in sedimentary rock, where biomass would have deposited. Taken together, this evidence made it seem like an open-and-shut case for the biological origin of these fuels.

It was the oil crisis of the 1970s, when the world's major economies faced sudden shortages as well as rapidly rising prices, leading to fears that the supply of oil might be running out (although, in hindsight, much of the crisis was due to tensions in the Middle East), that stirred Gold's interest in the question of where fossil fuels came from and how plentiful – or not – they might be. But, coming at the subject from an astronomical background, he quickly arrived at a conclusion very different from that of the geologists. In fact, he always maintained that the failure of geologists to accept his theory was because of a kind of disciplinary blindness on their part. They were so focused on the small picture of terrestrial geology that they totally missed the big picture provided by astronomy.

This big picture was that hydrogen and carbon are among the most abundant elements in the universe. Hydrogen is the most common element and carbon the fourth most common. Based on this alone, one would expect that hydrocarbons, being molecular combinations of hydrogen and carbon, should exist in relatively large amounts, and studies of the solar system have confirmed this expectation. Huge quantities of methane have been detected in the atmospheres of Jupiter, Saturn, Uranus and Neptune. Vast lakes of it also cover the surface of Saturn's moon, Titan. Presumably, all these hydrocarbons throughout the solar system are of non-biological origin. Given this, it didn't make sense to Gold that Earth's hydrocarbons should be of biological origin. Why would hydrocarbons form one way on our planet, but a different way everywhere else in the solar system?

This was the basic rationale that led him to conclude that the oil and gas pumped out of the ground must be of non-biological origin. The way he put it, geologists were simply hopelessly unaware of the astronomical data relating to the cosmic abundance of hydrocarbons. He told the story of one geologist who challenged him following a presentation by asking, 'Dr Gold, how many other astronomers believe your theory that methane exists in other parts of the universe?' Gold noted, 'I had to tell him that there aren't any astronomers who don't believe it.'

Gold supplemented his astronomical argument with evidence from terrestrial geology. He pointed out that it was common to find large quantities of helium in oil and gas, even though decaying biological matter shouldn't produce this element. So, where was it coming from? The only plausible explanation, he believed, was that, as methane migrated upwards from the deep Earth, it was leaching helium from the surrounding rocks.

He also observed that oil and gas fields are often associated with earthquake-prone regions, such as Iran. To him, this suggested that fault lines deep within the Earth must be allowing hydrocarbons to rise to the surface in these areas. He even speculated that hydrocarbons trying to force their way through the mantle might be one of the underlying causes of earthquakes.

Then there was the curious phenomenon of spontaneously refilling oil wells. Drillers would sometimes think they had exhausted a well, only to find the oil gradually returning. Conventional geological theory struggled to explain why this happened, but Gold's theory actually predicted it should.

Gold referred to these arguments, taken together, as his deep-Earth gas theory. He laid out the case for it in a series of articles, beginning in 1977 with an op-ed piece in the *Wall Street Journal* (an

unusual publication in which to launch a scientific theory) and cul-
minating with the book *Power from the Earth*, published in 1987.

There was, however, a big gap in his theory. He had explained
why hydrocarbons might exist in vast quantities deep underground, but he hadn't accounted for why oil had so many biological
characteristics on a molecular level. And this, after all, was one of
the main arguments for the biological origin of fossil fuels. A drilling project in Sweden provided him with a possible answer.

The Swedish national power company, Vattenfall, had been
desperate to find new sources of energy for Sweden. So, when
Gold began making headlines with his deep-Earth gas theory,
promising that oil and gas might be found in far more places than
conventional theory predicted, its engineers decided that, if there
was even a slim chance that he was right, the possible pay-off was
worth the gamble.

In the mid-1980s, the company put up tens of millions of dollars to finance an exploratory drill in the Siljan Ring, an enormous
bed of granite located in central Sweden, where a large meteorite
had impacted about 376 million years ago. According to conventional theory, it was the last place in the world one should be
drilling, because no competent geologist would expect to find gas
in a non-sedimentary rock such as granite. But, according to Gold,
it was exactly the right place, because the impact of the meteorite
would have fractured the crust of the Earth, allowing hydrocarbons to migrate upwards.

Drilling began in 1987 and continued for five years, reaching
depths of over 22,000 feet, which is far deeper than most wells.
The need for deep drilling was suggested by Gold's theory on the
logic that, if methane was rising up from the mantle, the richest
concentrations of fuel might be very far down.

Unfortunately for the investors, the operation proved to be a
financial bust. About eighty barrels of oil were eventually brought
up, which was intriguing given that mainstream theory said there
really shouldn't be any oil there at all, but it wasn't enough to
make any money. In addition to the oil, however, the drilling

unexpectedly brought up something else: evidence of microbes living almost seven kilometres down, wedged inside the pores of the rock. This was an extraordinary find, because biologists had long assumed that microbes couldn't survive more than a few hundred metres beneath the surface of the Earth due to the heat and pressure at such depths.

Gold realized that the existence of these deep-Earth microbes could fill the gap in his theory, explaining why fossil fuels possess so many biological characteristics. He reasoned that, if microbes were living at great depths throughout the crust of the Earth, they might be feeding on the hydrocarbons that welled up from the mantle. As they ate, they would both contaminate the fuel and help to transform it into oil, giving it the biological features that researchers had observed. As Gold put it, it wasn't that geology had transformed life to produce oil, but rather that life was transforming geology to produce oil.

Never one to do things by half measures, Gold took his speculations one step further. Given the enormous size of the Earth's crust, he concluded that it, rather than the surface, had to be the primary home of life on the planet. He imagined the existence of a vast 'deep, hot biosphere' that stretched downwards for miles beneath our feet, teeming with hydrocarbon-eating microbes.

Gold then went further still and connected his deep-Earth gas hypothesis to the origin of life itself. Life requires energy, and throughout much of the twentieth century theorists had assumed that the original source of that energy, the spark that gave birth to life, must have been the sun. After all, it's up there in the sky, constantly warming our planet. It's the obvious source. But Gold observed that, if methane has been continually seeping upwards from below since the Earth formed, then it could have served as a supply of fuel that nourished the earliest primitive organisms. It's actually far easier, in terms of chemistry, to extract energy from methane than it is from sunlight. So, methane would have been a logical energy source for early life. In which case, life could have started deep underground rather than on the surface. Instead of fossil fuels forming from fossilized life, it could have been the other way around. These fuels could have given birth to life.

Gold also made the intriguing observation that, if life on Earth had begun deep underground, this raised the possibility that life might exist within other planets. As he noted during a 1997 meeting of the American Association for the Advancement of Science, 'Down there, the Earth doesn't have any particular advantage over any other planetary body.' Our solar system, he suggested, might be crawling with subsurface life just waiting for us to discover it.

He detailed these arguments in his 1998 book *The Deep Hot Biosphere*. Taken as a whole, they represented his grand reimagining of our planet's history. In this version of events, hydrocarbons had played the starring role. First, they had been trapped underground in vast amounts during the formation of the Earth, then they had brought life itself into existence, and now they supplied our civilization with a potentially endless source of energy.

Gold's theory struck at the heart of conventional geology. If he was right, then geologists were completely wrong about one of the most basic facts of their discipline: the origin of oil. Not only that, but instead of helping locate the fossil fuels that powered the world's economies, they had actually been hindering the search all these years, preventing new sources from being found in more locations. Naturally, geologists didn't take kindly to this suggestion.

They didn't disagree entirely with Gold, however. They actually fully accepted that it was possible for oil and gas to form non-biologically. There were, after all, known processes that allowed chemists to manufacture abiogenic oil, such as the Fischer–Tropsch process that Germany had used to produce oil during World War II. Geologists even conceded that *some* of the Earth's oil and gas may have formed non-biologically. It was the idea that *commercial quantities* of these fuels had formed this way within the Earth that they regarded as patently absurd.

Critiques of Gold's argument began with the formation of the

Earth. Geologists agreed that hydrocarbons existed in abundant quantities throughout the solar system. Yes, they noted, they actually had been aware of this! But they maintained that most of the Earth's primordial methane would have leaked into space when the planet was young and molten hot. In particular, when the Mars-sized body that was thought to have created the moon slammed into the young Earth and melted it, most of the methane must have cooked off.

But, even if large reserves of methane did remain in the mantle and deep crust of the Earth, geologists argued that it wouldn't migrate upwards through rock, as Gold claimed, because the rock at such depths isn't porous enough to allow this migration. Nor would the methane readily convert into more complex hydrocarbons, such as oil. It's just not that easy to transform methane into oil. The geologist Geoffrey Glasby insisted that this latter fact alone invalidated Gold's theory.

Despite the staunch opposition from geologists, interest in Gold's deep-Earth gas theory nevertheless endures. Gold died in 2004, a firm believer in it to the very end. With his death, the theory lost its most vocal champion, but a small group of believers continues to wage a David-and-Goliath battle against the geological orthodoxy.

Part of the reason for this is because of Gold himself. It's hard to dismiss him as a crackpot because, as already noted, he had an impressive track record. His wacky ideas had a habit of being vindicated. And, in fact, he managed to repeat this trick with his speculations about the existence of a deep, hot biosphere.

Microbiologists initially greeted his announcement that he had found microbes living seven kilometres underground with extreme scepticism. It just didn't seem possible that any organism could survive at such depths. The experts suspected that his samples had been contaminated by surface bacteria. But, within a few years, other researchers had vindicated his claim by finding indisputable evidence of deep subterranean microbes elsewhere, such as at the bottom of a South African gold mine and brought up during a deep-drilling project in the Columbia River basin.

These discoveries tied in with a broader revolution that was sweeping biology during the 1980s and 1990s, as researchers found 'extremophile' microbes living in all kinds of bizarre, seemingly hostile settings: deep-ocean volcanic vents, Antarctic ice and acidic sulphur springs. Life, scientists began to realize, is amazingly adaptable, able to flourish in conditions that were once assumed to be deadly. In fact, it's colonized every nook and cranny of the surface of the Earth and has apparently extended its domain far beneath it as well.

The discovery of this deep biosphere lent credibility to Gold's hypothesis that life may have originated underground. There's no consensus about where life did start, but many scientists regard a deep-Earth origin as quite plausible, even likely.

This discovery also added some support to his deep-Earth gas theory, because it raised the question of how these subterranean microbes were surviving. What were they feeding on down there? Gold's argument that they might be nourished by methane rising from the mantle didn't seem entirely crazy, given how crazy it was that these ultra-deep microbes existed at all. Microbiologists have suggested other ways the microbes might feed, such as by extracting hydrogen from rock or even by using radiation as an energy source, but, given how difficult it is for researchers to know exactly what's going on that deep beneath the surface, Gold's theory can't be ruled out entirely.

But there's a more basic reason why support for his theory endures. It's because we don't seem to be running out of oil and gas, despite repeated predictions that by this time we should have been. The most famous of these gloomy forecasts was made in the 1950s by the geophysicist Marion King Hubbert, who declared that US petroleum production would peak in 1970 and that, by the end of the twentieth century, the global supply would also begin to gradually but inexorably diminish.

The oil crisis of the 1970s convinced many that Hubbert was

correct, but we're now well into the twenty-first century and oil and gas production is still booming. The prediction of imminent scarcity seems to have been wrong. This hardly proves that we may actually have a million-year supply, as Gold envisioned, but his supporters do argue that the continuing profusion justifies scepticism towards the conventional geology that was generating such low estimates.

Popular fears of a looming shortage of fossil fuels have actually subsided to such an extent that they've been replaced by a new concern: that it's the abundance of oil, not its scarcity, which is the real problem. This is due to the overwhelming evidence of the environmental damage caused by the use of fossil fuels, and this suggests the possibility of a somewhat cruel irony. There may be a lot of oil and gas left, but we can't keep burning it. In fact, imagine if Gold was right and the Earth really did contain an inexhaustible supply of oil and gas; we would be in a situation analogous to Coleridge's ancient mariner: 'Water, water, everywhere, nor any drop to drink.' Or, in our case: oil, oil, everywhere, nor any drop to burn.

What if alien life exists on Earth?

You're very similar to a slime mould. You also have many of the characteristics of a slug, an earthworm, an intestinal parasite and pond algae. But don't take this personally. It's not your appearance; it's your cellular biochemistry, and you share this in common with all known life on Earth.

Evolution has fashioned life into a bewildering diversity of forms, ranging from infectious bacteria all the way up to oak trees and elephants, but these forms are somewhat superficial. Deep down, on the cellular level, all species are more alike than different. The cells of all living things 'speak' the same language, which is the genetic code hidden in DNA, and they all store energy in the same way, using ATP molecules. This uniformity suggests that all life on Earth must have come from the same source material – that life must have started just once, and the many varieties of species now populating the globe all trace back to that single origin event.

This is the consensus view of modern science, but in the early twenty-first century a small group of scientists led by physicist Paul Davies and philosopher/astrobiologist Carol Cleland challenged this orthodoxy. They didn't dispute that all *known* organisms are part of the same tree of life, but they made the case that there might be unknown organisms lurking around that may belong to separate, alternative trees of life. In other words, life may have started more than once on Earth, and the descendants of those other origins might still be among us.

The 'biosphere' denotes the global ecosystem of all living beings, and, in the words of Cleland, the members of those other trees of life would constitute a 'shadow biosphere' sharing our planet. Proponents of the hypothesis often refer to these beings, by extension, as 'shadow life', although sometimes they use other terms such as 'novel', 'non-standard', 'strange', or 'alien life' (alien because, while these alternative life forms may be native to the Earth, they would be profoundly foreign, from our point of view).

These names may make it sound as if the shadow biosphere is somehow paranormal – a phantom zone bordering our own reality. That's not the intention. Shadow life, if it exists, would be entirely physical and real. It would simply possess a biochemistry fundamentally different from our own. What it would look like, however, is unclear. Externally, it might appear to be just like standard, known forms of life, even while its inner workings would be utterly strange. Or perhaps it would be completely unlike anything biologists have seen before. We might struggle to even recognize it as being alive.

Most scientists aren't opposed, in principle, to the suggestion that life on Earth may have started more than once. After all, if life began by means of some natural process, which it presumably did, then this process could certainly have happened multiple times, perhaps in different geographical locations. That's a reasonable assumption.

But, despite this willingness to entertain the general notion of multiple origins, the scientific majority hasn't embraced the shadow-biosphere hypothesis for the simple reason that alternative forms of life have never been encountered. It's not as if biologists have been sitting around for the past 200 years, twiddling their thumbs. They've been actively peering into every nook and cranny of the natural world. If a second tree of life exists, you would think someone might have stumbled upon it by now.

It's a bit like the scientific objection to Bigfoot. Believers in this big hairy creature insist that a giant primate may have survived from ancient times to the present in the forests of the US Pacific Northwest. Zoologists concede that, in theory, this wouldn't be impossible, but they point out that no biological specimens of such a creature (no hair, bones or body) have ever been found. So, as the decades pass with no sign of Bigfoot, the continuing absence of evidence eventually leads to the conclusion that he doesn't exist. And so it is with the shadow biosphere. It would be nice to believe it exists, but the simple fact is that no one has ever produced solid evidence that anything like it is out there.

However, shadow-life advocates counter that this analogy is flawed. Why? Because, if alternative forms of life exist, they're probably microbial in size. The overwhelming majority of species on Earth are. In which case, they would have the entire vast microbial realm in which to hide. Current estimates suggest that as many as a trillion species of microbes might exist on the planet. Of these, only around 0.001 per cent have ever been studied by scientists. So, it wouldn't stretch credulity at all to imagine that some of the hundreds of millions of undiscovered microbial species might be exotic types of shadow life. Tracking down Bigfoot in a forest, by contrast, would be a piece of cake.

Finding these creatures would be particularly difficult if, on the outside, they resembled known microbes. We would never realize, by visual inspection, how peculiar they were. This is likely to be the case, since microorganisms often have fairly generic shapes. Archaea and bacteria, for example, are two distinct types of single-celled microbes that occupy entirely different domains on the tree of life. In their own way, they're more different from each other than you are from a mushroom. But, through a microscope, they look the same. Scientists have to examine them on a genetic level to tell them apart.

This raises the issue of the crudeness of the tools microbiologists have at their disposal, which increases the difficulty of finding any type of shadow life. Microscopes, as noted, may not help much in the search. Researchers can also try to culture or grow microbes in the lab to study them more closely, but doing so is a delicate, tricky process. It's estimated that fewer than one per cent of known microbes have ever been successfully cultured. Shadow-life microorganisms would probably be among the 99 per cent that have never been successfully cultured. There are also various tools to analyse microbial genetic material, but these tools are designed to work with normal DNA. They would be useless for finding organisms that lacked this.

Given the constraints of these tools, argue the shadow-biosphere advocates, it's easy to imagine that, if there was an unusual microscopic form of life floating around out there that looked vaguely similar to known microbes, but resisted being cultured, and didn't have standard DNA (if it had DNA at all), then it could have entirely escaped our attention.

Biologists, if pressed, might concede this could be the case. But there's another reason they doubt the existence of the shadow bio-sphere. Given the aggressiveness with which standard life has pushed into every corner of the globe, it seems likely that any rival form of life would simply have been wiped out. In other words, life may in fact have started on Earth more than once, but, due to the fierce competition for resources, presumably only the one variety of it now remains.

But, again, shadow-biosphere advocates come right back with a counterargument. They point out that there are ways in which an alternative life form could have survived to the present day. It could have found an out-of-the-way niche to exploit, where it wouldn't have competed directly with standard life, perhaps an extreme environment such as deep inside a hot volcanic vent or underground in a region of high radioactivity. Or perhaps it learned to feed on chemical resources that were unpalatable to standard life. If so, these odd life forms would simply have been left alone and allowed to establish an independent ecological system, side by side with us. Whatever the answer, it's possible to imagine ways in which shadow life could have survived. It could be out there.

Truth be told, most scientists are fairly open to these arguments in favour of the shadow biosphere – as long as they're offered up as purely speculative ideas. However, some advocates have taken the argument one step further, and this is where the hypothesis really boils up into controversy and meets stronger resistance, crossing over into territory that mainstream science considers outrageous. These advocates not only suggest that alternative life *might* exist, they say they may have identified a few examples of it. They've compiled a list of odd stuff that, they think, may be alien forms of life here on Earth. It's like a speculative menagerie of weird non-standard organisms. If even one of the things really is non-standard life, it would be one of the most momentous discoveries of modern science.

So, what's on the list? One of the items is something you might have seen with your own eyes if you've ever gone for a hike or drive through a desert. It's a dark, shiny substance called desert varnish that forms on rocks in arid regions. Hundreds of years ago, Native American tribes used to make petroglyphs by carving images in it, and when Charles Darwin stopped off in South America while sailing around the world on the HMS *Beagle*, he noticed the stuff glittering on rocks and puzzled over it.

Desert varnish can't have been produced by the rocks on which it forms, because it contains elements such as manganese and iron, which aren't in the rocks. It consists of many very thin layers, which suggests to geologists that it's been laid down by microbes. The problem is that, to date, despite a lot of investigation, no one has been able to identify a microbe that might actually be depositing it. Nor has anyone been able to figure out a non-biological way it could have formed. So, it remains a mystery. Cleland has argued that it could be an example of shadow life existing right out in the open, before our eyes.

Nanobacteria are another, far less visible, item on the list. You need an electron microscope to see these. They're tiny spherical particles that look somewhat like bacteria. Hence the name. But they're much smaller – almost ten times more compact than a typical bacterium. They've been found in a variety of places, including rocks, oil wells and (somewhat disturbingly) human tissue.

No one disputes that these nanobacteria exist. The controversy is over whether they're alive. According to conventional biology, they're far too small to be living, because all the necessary machinery of a cell (the genetic material, as well as parts to manufacture proteins and store energy) wouldn't fit inside them. Nevertheless, some researchers claim to have seen them reproduce, and to have found DNA inside them. In 1998, the Finnish biochemist Olavi Kajander sparked an uproar when he suggested not only that they might be alive, but that they might be responsible for various medical conditions, such as kidney stones, hardening of the arteries, arthritis, Alzheimer's and cancer. The fact that nanobacteria occupy this ambiguous status between living and non-living matter makes them, for shadow-biosphere advocates, a perfect candidate for non-standard life.

A third exhibit in the shadow-biosphere zoo is arsenic life. These strange creatures are microbes that, purportedly, are partially made out of the highly toxic element arsenic.

The biochemistry of all known life depends heavily on phosphorus. It's an essential part of both the DNA and ATP molecules. But, in its atomic structure, phosphorus is very similar to arsenic, which is why the latter is so deadly. It's hard for our bodies to distinguish between the two elements, and, when arsenic replaces the phosphorus in our cells, we die. Because the two elements are so similar, it occurred to researcher Felisa Wolfe-Simon that there might be organisms out there whose biochemistry is built around arsenic rather than phosphorus.

She went out looking for them, and, in 2010, she claimed to have found them living in California's arsenic-rich Mono Lake. When her findings were published online in the journal *Science*, they sent shockwaves through the scientific community, because these organisms, if real, would have contradicted what was believed to be a basic rule of life – that DNA is built out of phosphorus. Their existence also would have potentially proven the shadow-biosphere hypothesis correct.

The current consensus, however, is that neither desert varnish, nanobacteria nor arsenic microbes are valid examples of shadow life. Researchers are hopeful they'll find a non-biological explanation for desert varnish; nanobacteria are speculated to be some kind of cell fragments; while attempts to replicate Wolfe-Simon's results at Mono Lake were unsuccessful, leading to the conclusion that they must have been caused by inadvertent arsenic contamination in her experiment. Therefore, the orthodox position remains that there's just the one tree of life on Earth. We don't share our planet with a shadow biosphere.

Shadow-life advocates aren't giving up the search, though. They explain that part of what motivates them is their desire to answer a far larger question: does extraterrestrial life exist?

The fun way to find out would be to seek out and explore other worlds, *Star Trek* style. Sadly, that's beyond our current capabilities. Passively listening for signals from alien intelligences also hasn't turned up any evidence, and it restricts the search to technologically advanced civilizations. So, in the meantime, these advocates argue, it makes sense to look for clues on our own planet. We should try to better understand how life arose here. Did it emerge multiple times? If so, that would suggest it's prolific – that it's an inevitable by-product of geochemical processes and will form readily if given the chance. Confirming this would imply it probably exists elsewhere.

But, if life arose only once here – and, as long as we don't find any other trees of life, this is what we have to assume – that has somewhat grimmer implications. It means we can't rule out the possibility that it may be a highly unusual occurrence. Its presence may be a magnificent one-off event. A grand cosmic fluke. So, when we look up at the sky at night, there may be nothing else out there looking back at us. We might be entirely alone in the universe.

Weird became (partially) true: the Gaia hypothesis

When you exercise outside on a hot day, you sweat, and when you stand around in the freezing cold, you shiver. These reactions are your body's ways of maintaining a consistent internal temperature. No matter what conditions might be like outside, your body strives for an even 37 degrees Celsius within. Elaborate biological mechanisms have evolved to help it achieve this.

The Gaia hypothesis, introduced by James Lovelock in 1972, makes the case that, during the past 3.7 billion years, life has been actively regulating the environment of the Earth in a similar way, using various planet-wide feedback mechanisms to maintain conditions favourable for itself. In other words, the hypothesis imagines that life isn't a passive passenger on this planet. Rather, it's constantly shaping and altering the Earth for its own benefit.

To say that this hypothesis is controversial is an understatement. People either love it or hate it. For quite a while, its detractors within the scientific community prevailed, but, since the mid-1980s, aspects of the hypothesis have made a comeback and are now well accepted. Although, its supporters achieved this respectability in a curious way. They borrowed a tactic from the world of business: if your company name suffers from controversy, simply change the name. Likewise, Gaia, in its respectable form, doesn't go by that name. Instead, it's known as Earth system science.

Before he dreamed up Gaia, Lovelock had already established a reputation as a brilliant inventor of chemical devices. His most famous invention was the electron capture detector, which could sniff out minute amounts of chemicals in a gas. It was this device that revealed the dramatic build-up of ozone-destroying chlorofluorocarbons in the atmosphere, leading to a worldwide ban on their use.

Lovelock had a pronounced independent streak, which his success as an inventor allowed him to indulge by giving him the financial means to abandon both industry and academia and work for himself, out of a lab in his garden shed, in Wiltshire. He was fond of saying that most scientists nowadays are no more than slaves to their employers.

The Gaia hypothesis started taking shape in his mind in the mid-1960s, when NASA solicited his advice on how to figure out if life existed on Mars. The space agency had assumed, based on his reputation as an inventor, that he would design a life-detecting gadget they could send to Mars, but instead it occurred to him that NASA didn't need to send anything to the red planet. They could determine if Mars hosted life simply by analysing its atmosphere from here on Earth, because – and this was his great insight – life would inevitably reveal its presence by altering the chemical make-up of the atmosphere, keeping it in a state of disequilibrium.

This state, he observed, was what made Earth's atmosphere so obviously different from every other planet in the solar system. Our atmosphere is a mix of approximately 20 per cent oxygen and 80 per cent nitrogen. Oxygen, however, is a highly volatile element. Left to its own devices, it will soon react with other chemicals and disappear. But, instead, its levels stay constant on Earth because living organisms, such as plants, keep pumping out more of it. By contrast, the Martian atmosphere consists mostly of carbon dioxide, which is chemically non-reactive. It will remain exactly the same for millions or billions of years, in a state of calm equilibrium. This suggests there's no life there. (Although, in 2003, scientists detected occasional whiffs of methane in its atmosphere, which, being highly volatile, does teasingly hint at the possibility of life there.)

NASA didn't much appreciate Lovelock's observation because it undermined the whole rationale for sending a spacecraft to Mars. Nevertheless, it got him thinking about the relationship between life and the environment, and he soon took his insight one step further. He concluded that life wasn't just randomly modifying the environment; it seemed to be doing so in specific ways to maintain conditions favourable for itself. He noted, for instance, that over the past four billion years, the sun has grown significantly hotter as the amount of helium in it has increased. It's part of the natural process of stellar evolution (and the bad news is that it's still getting hotter, which in a few billion years will make the Earth uninhabitable). This should have caused the surface temperature of the Earth to rise correspondingly. But it hasn't. To the contrary, the Earth's temperature had remained relatively stable – at least, stable enough to maintain life. It never became a scorching inferno, like our neighbour Venus. Why?

Lovelock argued that it was because life on Earth was actively regulating the temperature by means of various bio-feedback mechanisms. One such example is the massive blooms of algae that form on the surface of the ocean as temperatures rise. They pull heat-trapping carbon dioxide out of the air, thereby lowering the global temperature. Similar methods keep a whole range of environmental factors, such as the alkalinity and pH of the ocean, within parameters that are optimum for life.

Lovelock didn't come up with all of this entirely on his own. The microbiologist Lynn Margulis collaborated with him to refine the biological details of the argument, while his neighbour, the author William Golding, coined the name Gaia, referring to the Greek goddess of the Earth.

The concept of Gaia fascinated the general public. After all, it was awe-inspiring to think that there was a force operating on a global scale that was safeguarding the interests of life. It was also comforting to imagine that a principle of harmony resided at the heart of nature, that all life on Earth (with the exception, perhaps, of human beings) functioned together for the common good.

Biologists, however, weren't quite as taken with it. They agreed that it was a nice fairy tale, but as science, they insisted, the hypothesis simply didn't work. There was no mechanism for it to do so. It required organisms to be doing things not for their own direct benefit, but rather for the benefit of the global ecosystem – and that, they argued, ran completely counter to evolution, which they considered to be the organizing principle of life.

Richard Dawkins and Stephen Jay Gould were among Gaia's most vocal critics. They hammered away on the point that evolution works by means of natural selection, which operates entirely on the selfish individual level and is entirely without compassion. Whatever traits best allow an organism to pass on its genes get selected. This happens ruthlessly, no matter the cost to anything else.

It's true, they conceded, that often an individual organism will have the best chance of passing on its genes by sacrificing its own interests for the good of the group it belongs to. But the good of the entire planet was too far removed and distant to exert any kind of selective pressure on an organism. As a result, there was no way for Gaia to emerge by means of natural selection. And, as far as the biologists were concerned, this meant there was no way for Gaia to emerge, period.

They also argued that it was meaningless to claim that Gaia maintained an environment favourable for life, because there is no best overall environment. Life tries to adapt itself to whatever setting it finds itself in. Some organisms live in Arctic ice; others live in thermal hot springs. Those conditions are optimum for them.

But what really drove the biologists to fits of rage was the Gaian idea that the Earth itself was a living organism. Lovelock frequently spoke about the Earth as if it were alive, because he said it was useful to compare the way organisms regulate their internal environment with the way similar forms of self-regulation could be found operating, on a far grander scale, across the Earth as a whole. When challenged on this point, he always insisted that he was speaking metaphorically. But many of his readers took him literally, and this horrified biologists. Making the planet out to be a living being seemed to them to be neo-pagan Earth worship, not science. The evolutionary biologist John Maynard Smith denounced Gaia as 'an evil religion', while the microbiologist John Postgate warned of 'hordes of militant Gaia activists enforcing some pseudoscientific idiocy on the community.'

So, the biologists did everything in their power to suppress the Gaian heresy, and they were quite effective. Lovelock complained that it became almost impossible to publish anything about the hypothesis in scientific journals.

All these criticisms remain perfectly valid, and Gaia remains something of a dirty word in biological circles. So how is it possible to say that the hypothesis is now considered to be partially true? It's because, while biologists could only see in it the looming spectre of pseudoscience, geophysicists had a very different reaction. They found the idea to be enormously intellectually stimulating. They weren't that concerned about whether Gaia played nice with evolution. Instead, they were focused on the big planet-wide picture of how the Earth functioned, and, on this scale, Gaia offered an exciting new perspective.

In the 1970s, the dominant paradigm of how the Earth's environment worked was that it was governed by a combination of two forces: geology (such as plate tectonics and volcanoes) and astronomy (the sun and asteroid impacts). The assumption was that these two forces were so overwhelmingly powerful that life could do little but passively adapt itself to them. But the Gaia hypothesis drew attention to what a powerful role life actually played in shaping Earth's environment. Once this was pointed out to them, geophysicists recognized that it was obviously true, although it hadn't been evident to them before. With the benefit of this new perspective, they began seeing all kinds of ways in which life had radically transformed the Earth: not just its oceans and atmosphere, but also its rocks and minerals, and perhaps even its crust as well.

Geologists now recognize that most of the types of minerals on Earth wouldn't exist if it wasn't for life. This is because they need oxygen to form, and it was life that put sufficient levels of oxygen in the atmosphere to allow this to happen. In fact, the Earth boasts far more types of minerals than any other planet in the solar system. Mineral diversity seems to be a signature of life.

Life also enormously accelerates the production of clay by causing rocks to erode more quickly, and this then plays numerous roles in geophysical processes, by trapping and burying huge amounts of carbon biomass. The clay also acts as a lubricant, softening and hydrating the crust of the Earth, which facilitates the sliding of tectonic plates. Some geologists speculate that, without life, the process of plate tectonics (and therefore the movement of the continents) might have shut down long ago.

So, geophysicists came to realize that life hadn't merely adapted to geology; it had altered it as much as it had been altered by it. The two had co-evolved together. By the mid-1980s, this Gaia-inspired study of the interlocking series of relationships between the geosphere and biosphere had developed into the branch of geophysics known as Earth system science.

Admittedly, Earth system science is a weaker version of Gaia. It makes no claim that life *deliberately* maintains conditions favourable for itself. Sticklers might argue that, for this reason, it's not really Gaia. But Lovelock certainly viewed the two as being equivalent, and many practitioners of Earth system science have been happy to acknowledge and defend the influence of the Gaia hypothesis on their work.

So, this is the compromise that's been worked out. According to biologists, Gaia is dead, and they take credit for killing it. But if you talk to geophysicists, the hypothesis is still going strong. In fact, it serves as a fundamental paradigm in their discipline. They just call it Earth system science rather than Gaia.

What if we've already found extraterrestrial life?

Does life exist elsewhere in the universe, or is the Earth its only home? This is one of the great questions people have pondered for centuries. Making contact with an extraterrestrial civilization would provide the most satisfying answer to this mystery, but, failing that, many scientists would settle for just finding plain old microbes on another planet. This discovery, if it were made, could address basic questions about the place of life in the universe, such as whether it's a one-in-a-trillion chance event, or if it tends to arise wherever possible. It could also enormously advance biological knowledge by giving us something to compare Earth-based life against. For these reasons, the search for extraterrestrial life has always been a major focus of space agencies like NASA.

But, according to one theory, the question of whether life exists elsewhere has already been definitively answered, in the affirmative. This theory has nothing to do with little green men who ride around in flying saucers abducting people and occasionally creating crop circles. Instead, it focuses on the two American Viking landers that, in 1976, became the first crafts to land successfully on Mars. They carried equipment designed to test for the presence of microbial life in the Martian soil. According to NASA's official statements, and the consensus view of most scientists, the landers found no conclusive evidence of such life.

Dr Gilbert Levin, however, strongly disagrees. For several decades, he's been waging a campaign to convince the scientific community that the landers actually did conclusively detect life, but that NASA, for various reasons, has been unwilling to admit this. Levin is in a position to address this issue authoritatively as he was one of the researchers who designed the life-detection equipment carried by the probes.

Levin got his start as a sanitary engineer. A sewage specialist. This may seem far afield from the world of NASA, and it was, but it meant that he spent a lot of time thinking about microbes because one of his first job responsibilities was to test things such as swimming pools for bacterial contamination. Back in the early 1950s, the way of doing this was time-consuming, taking several days to complete. Frustrated by this slowness, Levin invented a faster way to do it. He called it his Gulliver test, since, like Jonathan Swift's Gulliver, it found tiny beings.

His invention exploited the fact that all microbes need to eat and excrete. They take in nutrients, process them and then expel them as gaseous waste. Levin realized that it would be possible to detect the presence of microbes by testing whether a liquid nutrient was being converted into a gas. He did this by lightly lacing a nutrient broth with a radioactive isotope, and then squirting the broth onto or into a sample of whatever needed to be tested. If there were microbes present, they would eat the nutrients and expel them as a gas, and, because the nutrients had been radioactively labelled, a Geiger counter would detect the atmosphere above the sample growing more radioactive – a sure sign of metabolic activity, and therefore of microbes.

His Gulliver device worked like a charm. It sniffed out bacterial contamination in mere minutes or hours, rather than days. It was also exquisitely sensitive, able to detect even the slightest contamination.

In 1954, Levin heard that NASA was looking for equipment capable of detecting life on Mars, so he submitted his invention and, to his delight, and against fierce competition, it was eventually chosen. It took many more years to make it lightweight and compact enough to fit on a spacecraft. NASA also renamed it the 'labelled response' (LR) test, because that sounded more scientific. But, when the two Viking landers successfully descended onto the surface of Mars in 1976, Levin's test was part of their on-board package of biological experiments designed to look for extraterrestrial life.

Before the missions left Earth, the NASA scientists had established strict criteria about what would count as the successful detection of life. They decided that, if a biological experiment generated a positive result, it would then need to be validated by a second control experiment, in which the same test was performed on Martian soil that had been sterilized by heating it to 160 degrees Celsius for three hours. If the sterile sample produced no response, this would be interpreted as compelling evidence that a biological agent had caused the positive result in the first experiment and that life had been detected.

A few days after landing, the first Viking lander shovelled a sample of soil into the test chamber. Levin's automated equipment then squirted it with radiated nutrient broth and everyone back at NASA waited with baited breath to see what would happen. Soon, the Geiger counter recorded a rapidly rising level of radiation in the chamber. It was an unambiguously positive response. But, next, the control experiment had to be run. In another chamber, Martian soil was sterilized and then tested. This time, the Geiger counter recorded no change in the atmospheric radiation level. The conclusion seemed clear. The pre-mission criteria had been met. Life had been detected.

The second Viking lander, which carried identical equipment, had descended onto the Martian surface 4,000 miles away. When it subsequently ran the same sequence of experiments, it produced the same data. With these results in hand, Levin and the other NASA scientists began popping champagne to celebrate. It seemed like a momentous occasion in the history of science. The Earth was no longer the only known home to life in the universe.

But the celebration was to be short-lived. A few days later, the NASA scientists changed their minds and decided that no life had been detected, after all. The problem was that the other on-board experiments had produced results far more ambiguous than Levin's LR test.

The Viking landers carried two other life-detection experiments on board. The gas-exchange experiment looked for possible metabolic activity by measuring whether Martian soil, when wetted, would produce oxygen. The tests indicated that it did, but so rapidly that the response seemed more chemical than biological. And, when sterilized, the Martian soil still produced oxygen. This suggested no life.

Then there was the pyrolytic-release test. This measured whether anything in Martian soil would respond to artificial sunlight by absorbing radioactively tagged carbon from the air. If it did, this would indicate the possible synthesis of organic compounds by microbes. The equipment measured a small positive response, which was intriguing, but the designer of the test, Norman Horowitz, eventually decided that it just wasn't enough of a response to indicate life. Perhaps a peculiarity of the soil had caused the result.

The really damning result, however, came from a fourth experiment, called the gas chromatograph–mass spectrometer (GCMS), designed not to test directly for life, but rather for the presence of organic compounds, which are the carbon-based building blocks out of which all known living organisms are fashioned. Its results came back entirely negative, which surprised everyone. The assumption had been that at least a few organic compounds should have been present in the soil. But the GCMS indicated there were none at all.

The larger context also had to be taken into consideration. Mars just didn't seem like the kind of place that could support life. Temperatures there were well below freezing, the atmosphere offered no protection against ultraviolet radiation, and the environment was bone dry.

The positive results of Levin's LR test were therefore eventually called into question. Yes, something in the Martian soil had definitely caused the liquid broth to transform into a gas, but many of the NASA researchers felt that the response had occurred too quickly to be biological. They hypothesized that, if Martian soil contained a chemical such as hydrogen peroxide, this could have produced the observed reaction.

Given all these facts, the conclusion seemed disappointing, but unavoidable. There was no life on Mars. Gerald Soffen, the chief Viking scientist, announced this to the public at a press conference in November 1976, and it's remained the official stance ever since.

At first, Levin toed the party line. He sat quietly during the NASA press conference, even as Jim Martin, the Viking mission manager, elbowed him in the ribs and whispered in his ear, 'Damn it, Gil, stand up there and say you detected life!'

But, as the years passed, his dissatisfaction grew. He didn't think it right that people were being told that Mars was a lifeless planet when, so he believed, his experiment had clearly indicated otherwise. He also felt that the no-life verdict had led to a loss of public interest in Mars. In 1997, he finally went public with his dissent, declaring outright, 'the Viking LR experiment detected living microorganisms in the soil of Mars.' Ever since then, he's been a vocal thorn in the side of NASA.

Levin raises a series of technical issues to cast doubt on the no-life theory. First and foremost, he insists that a chemical agent couldn't have produced the results shown by his LR test. After all, heating the soil to 160 degrees Celsius stopped the reaction. This suggests that the heat killed whatever organism was producing the gas. Most chemicals, on the other hand, wouldn't have been affected by that temperature. Hydrogen peroxide, which the official explanation attributed the positive result to, certainly wouldn't have been. That was the entire point of heating the soil – to differentiate between a biological and chemical agent.

In 2008, the *Phoenix* lander did find perchlorate in the Martian soil. Like hydrogen peroxide, it's a powerful oxidant that could have produced a positive result. But Levin notes that perchlorate also wouldn't break down at 160 degrees Celsius. Therefore, its presence doesn't rule out the possibility of microbial life.

Levin also notes that it was the GCMS that tipped the balance in favour of the no-life conclusion by failing to find organic compounds, but subsequent experimentation on Earth has revealed that the GCMS had serious shortcomings. It failed to find organic compounds in both Chile's Atacama Desert and in Antarctic soil, even though they were certainly there. And, in 2012, NASA's *Curiosity* rover did detect organic compounds in Martian soil, further calling into question the GCMS results.

Levin has even suggested that there may be visual evidence of life on Mars. As early as 1978, he pointed out that some of the pictures taken by the Viking landers appeared to show 'greenish patches' on Martian rocks. A few of these patches, he claimed, shifted position throughout the Martian year. Sceptics have dismissed this as just a trick of the light, but he maintains that there may be some kind of bacterial substance growing in plain sight on the Martian rocks.

He also offers a broader argument, which focuses on the hardiness of life. Scientists once viewed life as being fragile, able to survive in only a limited range of environmental conditions. Given this belief, it wasn't surprising that NASA scientists in the 1970s concluded that life didn't exist on Mars – at least, not in the places where the Viking landers looked for it. But, since the 1970s, the scientific understanding of life's toughness has changed dramatically. Researchers now realize that life exists just about everywhere on Earth. They've found microbes surviving in the coldest parts of Antarctica, high up in the stratosphere, at the lowest depths of the oceans and even kilometres deep beneath the ground. Life, we now know, possesses an amazing ability to thrive in even the most extreme conditions.

Researchers have also identified meteorites on Earth as having come from Mars, which means that Earth and Mars aren't isolated from each other. They've been exchanging geological material for billions of years – the swapping of material presumably going both ways – thanks to asteroid impacts that blast rocks into space, allowing them to drift from one planet to the other. And, since microbes can survive inside rocks, we have to assume that Earth microbes must have long ago made their way to Mars.

Given this, Levin argues, it would be extraordinary if life *didn't* exist on Mars. In fact, a sterile Mars would radically contradict everything we've learned in the past few decades about life's hardiness.

Levin has attracted a small group of scientists to his cause. In 2015, he published an article listing fourteen of them who were willing to declare their belief that his LR experiment found Martian life. He also listed a further fifteen who think his test *may* have detected life. In this latter column were several prominent scientists, including the physicist Paul Davies and the geologist Robert Hazen.

One of Levin's most enthusiastic supporters is the Argentinian neurobiologist Mario Crocco, who in 2007 proposed that the life form (supposedly) found by the Viking LR test should be given the scientific name *Gillevinia straata*, in honour of Levin.

The majority of the scientific community, however, remains unconvinced, and many of Levin's former colleagues at NASA wish he would just shut up. In 2000, Norman Horowitz, designer of the Viking pyrolytic-release experiment, vented to a *Washington Post* reporter that, 'Every time [Levin] opens his mouth about Mars, he makes a fool of himself.'

This isn't to say that most scientists think Mars is lifeless. Far from it. There's a popular theory that life may be found on Mars in isolated 'oases', or pockets of liquid water underground. But the general feeling is that the Viking results were too ambiguous to offer any concrete proof for the existence of life on the planet.

Of course, the entire controversy could be settled definitively with more missions to Mars designed specifically to look for life, and Levin has been eager for this to happen. He advocates putting a video-recording microscope on a lander so that researchers could visually check to see if there's anything tiny wriggling around in the Martian soil.

NASA, on the other hand, seems to be in no hurry to settle the debate. It was heavily criticized for having created the Martian-life controversy by sending the experiments to Mars before scientists understood the environment there well enough to interpret the results meaningfully. So, the agency's strategy is now to proceed slow and steady. The *Curiosity* rover, for instance, which has been on Mars since 2012, hasn't done anything to look directly for life. It's only looked for things that might make life possible, such as the presence of water. Levin has complained that, despite its name, the rover has shown a distinct lack of curiosity.

NASA promises that a future mission to the planet will collect samples of Martian soil, but only so they can be brought back to Earth at a later date and examined at leisure by researchers, and no one knows when that later date might be. It could be decades in the future. Until then, the controversy will persist.

CHAPTER FOUR

The Rise of the Psychedelic Ape

We've seen that life established itself on Earth at least 3.7 billion years ago. The earliest form of it about which there's clear evidence consisted of little more than thin films of single-celled organisms coating the surface of rocks near hydrothermal vents. But one of the most important traits of life is that, over time, it evolves. So, let's watch in fast-forward as the ancestors of these primitive first cells transform.

For several billion years, the Earth belongs exclusively to the single-celled microbes. They fill the oceans, feeding at first on free-floating organic compounds before learning how to capture the energy of the sun. Eventually, they figure out ways to combine together into multicellular organisms. But it's not until 550 million years ago that abruptly they explode into a bewildering diversity of animal forms. They're still in the ocean, but gradually, after another hundred million years, they crawl out onto the land – first as plants, then amphibious creatures, and finally as reptiles that roam far and wide. These reptiles grow into the dinosaurs that rule the Earth for 150 million years, before most of them suffer a cataclysmic end, sixty-five million years ago.

Meanwhile, more humble animals have emerged alongside the dinosaurs: small mammals. And, as the dinosaurs exit the stage, these mammals rise to take their place. One of them in particular is quite peculiar. It's a small squirrel-like creature that lives in trees.

Fast-forward another forty million years and the descendants of these squirrel-like creatures are still living in trees, but they've now grown bigger and are recognizable as primates. Another fifteen million years pass, and some of them are on the ground. And then, about eight million years ago, a remarkable thing happens. A few of them begin to stand on two legs and walk upright.

These, of course, are our direct ancestors: the first hominins. It's the puzzle of how exactly these creatures appeared and transformed into us that we'll now examine.

Palaeoanthropology is the branch of anthropology that specializes in addressing this question. These are the researchers who dig up early human fossils and try to tease out clues from them about our evolution. However, these scientists constantly have to fight off the encroachments of outsiders, because how the human species emerged is a topic about which many people have a very strong opinion. It's these outsiders – some of them amateurs and others from disciplines such as marine biology or genetics – who tend to produce the weirdest theories about our origins.

What if the dinosaurs died in a nuclear war?

Sixty-five million years ago, something killed the dinosaurs. It's lucky for us that this happened, because, as long as those mighty reptilian predators ruled the Earth, the small mammals that lived alongside them never stood a chance of developing to their full potential. They were too busy trying to avoid becoming dinner. And, if the mammals had remained second-tier residents of the Earth, we would never have come into existence. So, what was it that happened to the dinosaurs? What brought them down?

The leading theory, proposed in 1980 by the father-and-son team of Luis and Walter Alvarez, is that a giant asteroid strike wiped them out. But many other scenarios have been suggested over the years, and at the more unusual end of the scale is the idea that the dinosaurs may have killed themselves in a nuclear war . . .

The theory that the dinosaurs may have suffered self-inflicted nuclear annihilation occurred, believe it or not, to two different people at around the same time. They were John C. McLoughlin, a writer, artist and author of several popular works about evolutionary biology, and Mike Magee, a retired Yorkshire chemist. McLoughlin came up with the idea first, describing it in a 1984 article in *Animal Kingdom* magazine, while Magee presented his version slightly less than a decade later in a self-published book titled *Who Lies Sleeping: The Dinosaur Heritage and the Extinction of Man.*

Although McLoughlin thought up the concept first, there's no evidence that Magee was aware of this. It really does seem to be the case that both men independently had the same eureka moment, and the arguments they made were very similar. So similar that, for our purposes, we'll treat them as one joint conjecture, which we'll call the atomic-dino hypothesis, since neither of them ever coined a term for it.

The phenomenon of two people coming up with the same idea at around the same time is referred to as 'multiple independent discovery'. A well-known example of it from the history of science is the near-simultaneous but separate discovery of the concept of calculus by Isaac Newton and Gottfried Wilhelm Leibniz in the late seventeenth century. Usually, such an occurrence is a sign that, culturally, something is in the air that has made an idea ripe for being dreamed up, and this was definitely the case for the idea of dino nuclear war.

In 1983, a group of researchers, including Carl Sagan and Paul Ehrlich, had published an article in *Science* detailing how a nuclear war would throw huge amounts of sunlight-blocking dust into the atmosphere, thereby triggering a 'nuclear winter' that would plunge the surface of the Earth into a deep freeze, making it uninhabitable for years. Before the 1980s, the prevailing wisdom had been that such a war would kill a lot of people, but that a substantial number would survive. The *Science* article sparked a growing awareness, bordering on mass panic, that a full-scale atomic conflict between the Cold War superpowers could actually bring about our extinction as a species.

This mood of nuclear doom and gloom evidently set McLoughlin and Magee thinking, leading them both to connect the same dots: if a nuclear war could cause a planet-wide extinction, then what if that was what killed the dinosaurs?

Both authors were fully aware of how outrageous such a suggestion sounded. McLoughlin, for instance, emphasized that it wasn't an idea he actually believed in. He described it as a stray thought that popped into his head late at night that he was unable to get out of his mind. He blamed it on listening to the music of Bartók, that 'mad Hungarian'. Magee committed himself more fully to the idea, though he also gave himself an out, blaming his pursuit of the concept on the ineffectiveness of the local cider at dulling his fervid imagination. Despite such disclaimers, both men nevertheless went ahead and made the argument in print.

In order for the dinosaurs to have died in a nuclear war, some of them must have been smart enough to build nuclear bombs. There had to have been a dinosaur species that evolved advanced, tool-making intelligence. This is the first and central pillar of the atomic-dino hypothesis. It asserts that there's no obvious reason why this couldn't have happened.

This may sound crazy, but let's consider what's required in order for this kind of intelligence to evolve. To answer this question, we only have one example to draw from: our own species. By studying our ancestors, anthropologists have identified some factors that they believe played a crucial role in causing early humans to develop large brains (and, eventually, sophisticated technologies).

At the top of the list are a variety of anatomical features that include opposable thumbs, walking upright and binocular vision (two eyes, with overlapping fields of view, giving depth perception). All of these, in combination, allowed our ancestors to use their hands to manipulate tools, which promoted intelligence.

Living in social groups also placed enormous selective pressure on brain growth, because navigating these relationships is a highly demanding mental task. Finally, and perhaps most unexpectedly, there was our diet. Our ancestors ate meat. It turns out that a species can possess all the previous attributes, but if it lacks an energy-rich diet of the kind that meat most readily provides, it's very unlikely to ever develop a large brain, because these need a lot of fuel.

So, early humans were group-living, carnivorous bipeds with opposable thumbs and binocular vision. These were the qualities, anthropologists believe, that predisposed them to get smart. Did any dinosaurs share these traits? Yes! McLoughlin offered the example of the deinonychus – the bipedal, pack-hunting predator featured in the movie *Jurassic Park*, where they were referred to as velociraptors. (The film-makers knew they were using the wrong name, but later argued that velociraptors just sounded better than deinonychus.) These creatures didn't have opposable thumbs per se, but they had powerful grasping claws, which is the source of their name. Deinonychus means 'terrible claw'. They definitely had all the other traits.

In fact, there were quite a few dinosaur species with similar characteristics, like the stenonychosaurus. So, based on this list of features, it would be reasonable to predict that at least one dinosaur species should have evolved higher intelligence.

And yet, as far as palaeontologists know, this didn't happen. The dinosaurs seemed to stand on the cusp of developing big brains, but they didn't take that next fateful step. Could something else have blocked their progress? Some kind of large-scale environmental condition?

One possibility is that slightly lower atmospheric oxygen levels during the Cretaceous period might have inhibited brain growth, because brains are oxygen-hungry organs. Or perhaps the world of the dinosaurs was too full of dangerous predators. Big brains require more training, which means that infants need to remain dependent on their parents for longer. Human children take well over a decade to mature. In the savage world of the dinosaurs, such a prolonged period of helplessness could have been lethal. Or an even simpler explanation may be that the dinosaurs simply ran out of time. Several palaeontologists have speculated that, given a few more million years, intelligent dinosaurs might have evolved.

The thing is, though, none of these obstacles seem insurmountable. The lack of time, in particular, is questionable. The dinosaurs had 150 million years. How much more did they need?

The atomic-dino hypothesis suggests that, instead of trying to explain why dinosaurs didn't evolve intelligence despite seeming to possess the necessary prerequisites for it, we should consider the possibility that one species actually did, and we just don't know about it yet.

This leads to the second pillar of the atomic-dino hypothesis, which makes the argument that the fossil record isn't complete enough to allow us to say with absolute certainty that brainy dinosaurs, and by extension dino civilizations, never existed. There's enough wiggle room to create plausible doubt.

Palaeontologists themselves will certainly concede that the fossil record is an imperfect archive. Not everything gets preserved. Using the fossil record to reconstruct the past is a bit like trying to piece together the plot of a movie from a few still frames. You're never sure how much you're missing. Characters and entire plotlines may be omitted.

Consider this – the entirety of the fossil record documenting the past six million years of chimpanzee evolution consists of three teeth found in 2005. Three teeth! If, for some reason, chimpanzees had gone extinct a thousand years ago, we probably wouldn't even be aware that such a species ever existed. The reason for this paucity is because certain environments, such as jungles, don't preserve bones well. What gets recorded in the fossil record depends entirely on the chance circumstances of where an animal happened to die.

Plus, the further back in time you go, the worse the fossil record gets. If there had been a dinosaur species that experienced rapid brain growth during the final million years of the late Cretaceous period, but it lived for most of that time in a jungle, it's plausible that we could be entirely ignorant of it.

But wouldn't an entire civilization leave behind evidence of its existence? Perhaps. Or perhaps not. What would remain of our own civilization in sixty-five million years if we were foolish enough to wipe ourselves out, or were unlucky enough to be wiped out by an environmental catastrophe?

That's a long time to erase evidence. Earthquakes, floods, tornadoes and hurricanes would rip apart our cities. Sun, wind and rain would erode the rubble. Rising sea levels would drown it. Glaciers would descend from the poles, grinding everything in their path down to a fine dust. Over such a vast expanse of time entire continents would shift and new mountain ranges would rise up. Though it may hurt our vanity to admit, the mark we've left on the planet may not be as permanent as we like to think. The Earth could eventually forget us.

There's an even more unsettling possibility, which forms the final pillar of the atomic-dino hypothesis. It's possible that evidence of this former dinosaur civilization did survive, but we're failing to recognize it for what it is. We're not seeing what's right before our eyes.

Let's look at the geological evidence that led researchers to conclude that an asteroid killed the dinosaurs. They found a sediment layer rich in rare elements and heavy metals, as well as chemical signs of massive fires and acid rain at the end of the Cretaceous. It all suggested that some kind of cataclysm occurred at that time, which certainly could have been caused by an asteroid strike. But couldn't these also be the telltale signs of industrial pollution and nuclear war? The explosion of atomic weapons followed by a nuclear winter could produce effects very similar to an asteroid strike.

There are other clues. It turns out that the dinosaurs didn't die out abruptly. The fossil record indicates a gradual decline that began around a million years before the end. Something was killing them off before they disappeared completely. It didn't happen in one moment.

Intriguingly, a similar phenomenon has been occurring during the past 100,000 years of our own era. As our ancestors began spreading out across the globe, equipped with the full mental capacity of modern humans, it seems increasingly clear that they systematically drove to extinction almost every large-sized species they encountered. In North America, they killed mammoths and sabre-toothed tigers; in Australia, the giant kangaroos; in Europe, it was their closest competitor, the Neanderthals. Scholars refer to this mass slaughter as the Anthropocene extinction.

The point is that a slow-moving mass extinction, such as we see in the late Cretaceous, could be evidence of the rise of an intelligent super-predator – a species of dinosaur that far out-performed those around it.

And there's one more clue in this puzzle, because not all species began going extinct towards the end of the Cretaceous. There were some dinosaurs that survived in substantial numbers right up to the very end. These were the ceratopsians, a group of large herbivorous horned dinosaurs that included the triceratops. What kept them alive, if everything else was dying out?

Again, the history of our own species might offer a possible explanation, because the Anthropocene extinction hasn't affected all species equally. To the contrary, some species have benefitted greatly from our presence. In particular, various large herbivores, like cows and sheep, have flourished in staggering numbers, even as most other species have been dropping off like flies.

What if something similar was occurring in the late Cretaceous? What if the ceratopsians survived in such high numbers because a brainy species of dinosaur was farming them for meat? What if the intelligent dinosaurs had a taste for triceratops burgers?

Brainy nuke-building dinosaurs? Triceratops burgers? Admittedly, all this might be a hard pill to swallow. Certainly, there are no palaeontologists who take any of it seriously.

Yes, the effects of an asteroid impact might vaguely resemble those of a nuclear war from a vantage point of sixty-five million years. But surely some fragment of dinosaur technology would have survived if it ever existed. A piece of a dino-gun or a dino-engine. A dino-nuke! But there's nothing.

There's also the Chicxulub impact crater, off the coast of Central America, which is hard to explain away. It was identified in the early 1990s as the site where the dinosaur-killing asteroid probably landed. In other words, scientists are pretty sure that something big hit the Earth, and, if this is the case, there's no need to conjure up a nuclear war in addition to account for the extinction of the dinosaurs.

The atomic-dino hypothesis does, however, touch on a larger issue, which may be a reason for not dismissing it as entirely frivolous. It's the question of why we're the only species to have ever evolved advanced, tool-wielding intelligence. One would think that this is such a useful ability to possess that evolution, in its constant search for competitive advantage, would have hit upon it more than once. Other complex traits, such as flight and eye lenses, have evolved independently multiple times. And yet, we're apparently the only species that has ever managed the trick of developing brains with the power to reason – not only in the history of the Earth, but, as far we know, in the entire universe. Gazing up at the stars, we find no evidence of other civilizations out there. Why is this?

The widely assumed answer is that it must be highly improbable for intelligence of our kind to evolve. It must require such a rare confluence of circumstances for this to happen that it essentially never does, with ourselves being the great exception. This makes us quite extraordinary.

The atomic-dino hypothesis offers a gloomier possibility. It suggests that our intelligence is actually not that unique, and that clever species like ourselves may, in fact, have evolved before, both on our own planet and elsewhere. The problem, however, is that technological civilizations, when they do emerge, are inherently unstable. They tend to wipe themselves out in short order, leaving behind little evidence of their existence. This is why the stars are silent, because everyone out there who was like us has destroyed themselves.

Such an idea may make us uncomfortable. It's certainly easier to believe that we're one-of-a-kind, but the atomic-dino hypothesis whispers a contrary warning: our advanced technology doesn't make us as special as we like to think. In fact, if we're not careful, it may cause us to go the way of the dinosaurs.

What if our ancestors were aquatic apes?

Imagine our ancestors ten million years ago. They may have looked a bit like gibbons – covered in fur, hanging from branches by their arms, running around on all fours between trees. Somehow, those primitive creatures transformed into us. Some of us are still rather hairy, but otherwise we're dramatically different from those earlier primates. How did this transformation happen?

According to anthropologists, the change occurred during the millions of years our ancestors spent wandering the African landscape, migrating through woodlands and across dusty savannahs, until finally, after many generations had passed, their bodies came to look like ours. But, for over half a century, the aquatic-ape theory has made the case for a radically different and far stranger version of our history. It argues that our ancestors didn't remain entirely on land, but instead took to the water for over a million years, swimming among the waves like dolphins, before eventually resuming a terrestrial existence. It was this ocean interlude, the theory says, that transformed them from ape-like creatures into beings resembling ourselves.

The father of the aquatic-ape theory was British marine biologist Alister Hardy. The idea came to him in 1929, while he was reading in his study. He was making his way through *Man's Place Among the Mammals* by the naturalist Frederic Wood Jones when he came across a passage discussing the mystery of why humans have such a thick layer of fat beneath their skin. This made Hardy think of the various marine creatures he had observed on a recent trip to Antarctica. Many of them possessed a thick layer of blubber, and the parallel between this and human fat intrigued him. It was then that he had his flash of inspiration. A vision popped into his head of our ancestors living as aquatic apes. Or, more precisely, as semi-aquatic apes. He pictured them dwelling along the African coast and diving into the water after fish, in the manner of penguins or seals, with the water streaming over their thin covering of fur.

Hardy kept his aquatic speculation to himself for thirty long years, fearing his colleagues would consider the notion absurd. He only revealed it to the world in 1960, at a meeting of the British Sub-Aqua Club. And it turned out he was right. His colleagues did think his theory was absurd. Faced with this rejection, Hardy moved on to another project: the search for a biological basis for telepathy. He had a fondness for against-the-mainstream ideas.

The affair of the aquatic ape might have ended there, barely having caused a ripple in academia, but, a decade later, out of the blue, it found an unlikely new champion: a fifty-two-year-old Welsh scriptwriter named Elaine Morgan. She had no scientific training at all, but she loved reading about science, particularly human evolution, and when she came across a reference to Hardy's theory she said it was 'as if the whole evolutionary landscape had been transformed by a blinding flash of light.'

Morgan became an instant convert and hard-core believer. She was so impressed that, despite her lack of a scientific background, she decided to launch a one-woman campaign to promote the aquatic ape. Her first book on the subject, *The Descent of Woman*, became an international bestseller and was eventually translated into twenty-five languages.

Morgan consequently became the mother of the aquatic ape, and it's really to her that the theory owes its cultural prominence because she never gave up campaigning for it. She almost single-handedly raised it from obscurity, attracting to it a larger following than almost any other unorthodox theory in modern science enjoys.

The question at the heart of the aquatic-ape theory is this: why are we so unique as a species, in terms of our physical appearance? Just think about it for a moment. We really do look somewhat odd. Even among our primate relatives, like gorillas and chimpanzees, we stand out like a sore thumb.

Take the fact that we walk upright, on two legs. That alone marks us out as strange. How many other creatures are bipedal? Kangaroos perhaps, though they hop rather than walk. And birds – but again, they don't walk, they fly. Why did it make evolutionary sense for almost all other terrestrial species to keep going around on all fours, but not us?

Then there's our lack of fur. Having a covering of fur is eminently practical. It protects against the sun during the day and keeps you warm at night. Why did early humans lose theirs, while the vast majority of other land mammals kept a nice thick coat?

Throughout most of the twentieth century, from the 1920s to the 1990s, the standard anthropological answer to the puzzle of our appearance was that living on the open savannah was responsible. The idea was that, after our ancestors left the jungle, they moved on to the open plains of Africa, where they started walking upright because this allowed them to see over the tall grasses to spot predators. Upright, they could also hold and throw weapons. They lost their fur in order to stay cool while running around in the hot sun.

In the 1990s, this explanation had to be revised. Researchers were able to use fossilized pollen to reconstruct how the African landscape had changed over time, and they discovered that savannahs weren't actually very common in Africa until two million years ago, by which time our ancestors were already hairless and walking upright. The proto-humans, instead, would have inhabited a landscape filled with scattered woodlands and lakes, and it's less obvious why bipedalism evolved in this environment. Researchers have speculated that standing upright may have allowed our ancestors to grab food from overhead branches. Or maybe it began as a way to walk along branches, like orangutans do. The loss of body hair still seems to have been a way of regulating internal temperature, though perhaps it also made early humans less susceptible to lice and other skin parasites.

Despite these revised explanations, the larger and accepted anthropological point of view has remained consistently the same, which is that the African landscape moulded our bodies into their current form.

Aquatic-ape advocates reject these explanations as nonsensical. That's the crux of the dispute between them and mainstream science. They don't believe these answers adequately account for our uniqueness. After all, they ask, whether our ancestors lived on the savannah or in woodlands, if it made so much sense to stand upright, why didn't any other creatures make the same evolutionary choice? There were plenty of other animals living in the same landscape, including other primates. Why did only our ancestors discover the benefits of walking on two legs? And, if losing our fur helped so much with heat regulation, or controlling parasites, why did the vast majority of other land animals keep theirs?

The aquatic-ape theory concludes, to the contrary, that our singular appearance can only be explained if our ancestors weren't exposed to the same landscape as all the other land mammals of Africa. Instead, they must have taken an evolutionary detour through an entirely different environment – the water. This then caused the development not only of our hairlessness and bipedalism but of a whole slew of other quirky anatomical attributes.

It was around five to seven million years ago, the theory contends, when a population of apes left their home in the jungles of Africa, journeyed to the coast and marched into the ocean. These apes became the founders of the human lineage.

It's not entirely clear why they would have chosen to adopt

an aquatic lifestyle. Perhaps it was because the water offered protection from predators such as big cats. Or they may not have made the move voluntarily. Climate change and flooding could have trapped some apes on a large island, forcing them to turn to marine resources for survival. Whatever the reason, the theory imagines that the apes continued a coastal existence for one or two million years, spending much of their time foraging in shallow water for shrimp and crabs, diving into deeper water for fish and sleeping at night on land. In this way, they gradually adapted to an aquatic environment before eventually returning fully to the land. Like the initial move into the water, the reason for this return isn't totally clear. In the case of the island scenario, climate change may have reconnected their habitat to the mainland, prompting them to migrate along the coast and into the African interior.

This aquatic episode, according to the theory, permanently changed the bodies of the proto-humans in many ways. At the top of the list, as already noted, was our upright posture. There's no environment on land that would have forced our ancestors to stand primarily on two legs. Other primate species, such as chimpanzees and baboons, walk on branches and grab fruit from overhead, and yet they continue to go around primarily on all fours.

If a group of apes had ventured into the water, however, they would have instinctively stood upright to keep their chests and heads dry. Modern-day chimps and gorillas do exactly this when wading across streams. As the ocean-going apes waded further out, the buoyancy of the water would have made it natural for them to rest their feet on the bottom while their heads bobbed above the waves.

Then there's our lack of body hair. In the water, this would have helped to reduce resistance as we glided through the water. Hairlessness is an adaptation seen in many marine mammals, such as dolphins and seals. Though, it would have made sense to keep some hair on top of our heads to prevent our scalps from getting burned as they poked out above the water.

Bipedalism and hairlessness, however, are just the tip of the iceberg. Aquatic-ape advocates have a long list of other ways that the marine environment altered our anatomical features.

Beneath our furless skin, for example, lies a relatively thick layer of subcutaneous fat – far more than most other mammals have. This is the feature that inspired Hardy to dream up the aquatic-ape theory in the first place. It doesn't seem to serve any obvious purpose on land, but in water it might have kept us warm and buoyant, much like the blubber found in whales, seals and penguins.

The protruding shape of our noses, which is so different from other primates, conveniently deflects water over and away from our nostrils when we dive in. We have a poor sense of smell compared to other land species, perhaps because there was no need to be able to smell while swimming. We also have voluntary control of our breath. That is, we can consciously regulate how much air we breathe in and how long we hold it in our lungs, which is why we can dive underwater for extended periods. Other land mammals can't do this as well as we can, and it's this ability to control our breath that allows us to speak. Our bodies even possess an instinctive diving response. When we jump into cold water, our heart rate and other metabolic processes automatically slow down, reducing our body's consumption of oxygen. It's an odd ability for a land mammal to have, but it would be entirely sensible for an ocean-going creature.

The list goes on and on, but the overall argument should be clear. The evolutionary choices made by our ancestors seem puzzling if they never left the land, because so often they changed in ways completely unlike the creatures around them. Aquatic-ape fans insist this must mean that our ancestors lived for a significant amount of time in the water.

For a long time, anthropologists tried to ignore the aquatic-ape theory – or the soggy-ape theory, as they sometimes like to call it. To them, its claims seemed to be so self-evidently absurd that they didn't require refutation. They assumed the public would recognize the silliness of the whole argument. As the years passed, however, it became clear that the general population wasn't having this reaction. In fact, a lot of people seemed to think the theory made pretty good sense. Increasingly vexed by the persistence of the soggy apes, anthropologists gradually swung into debunking mode.

Nowadays, if you dare broach the subject with an anthropologist, expect to get an earful in response, peppered with terms such as 'pseudoscience', 'claptrap' and 'rubbish'. Press further and you'll also get a long list of the reasons why they believe the theory to be wrong.

Topping this list is their conviction that the theory clumsily interjects an unnecessary extra step into the story of human evolution. We know present-day humans are terrestrial, as were our jungle ancestors. It's simplest, therefore, to assume that our species has always lived on land, so we only need to account for the evolution of each feature once. If we speculate that our ancestors went through an aquatic phase on their way to becoming human, we have to explain how features first evolved in a marine habitat, and then why they were retained following the return to land. In the technical jargon of science, this extra explanatory step is non-parsimonious.

They also strongly disagree that any of our features seem designed for aquatic living, insisting that all of these apparent 'marine adaptations' dissolve upon closer examination. Take hairlessness. Losing our hair wouldn't actually have given us any appreciable swimming advantage. There are, as we know, many semiaquatic mammals that are furry, such as otters.

Also, if our ancestors really had lived in a marine habitat, there are adaptations our species should have developed, but we never did. One random, but illustrative, example is internal testicles. All aquatic mammals keep theirs inside their body, because having these bits dangling in chilly water isn't good for reproductive fitness. Best to keep them inside the body, where it's warm. Our species' external testicles strongly suggest we've always been terrestrial.

Anthropologists also complain that much of the supposed evidence for the theory focuses on the evolution of soft body tissues, such as skin, hair and fat. The problem here is that these tissues aren't preserved in the fossil record, which makes it difficult to know exactly when in our history such features developed or under what conditions. Because of this ambiguity, it's easy to spin 'just-so' stories about the origin of these features, which is why anthropologists prefer to focus on the 'hard' evidence of fossils, which don't give any indication of an aquatic phase in our evolution.

Perhaps anthropologists anticipated that, once they had explained why the aquatic-ape theory was wrong, everyone would wise up and lose interest in it. This hasn't happened. Today, it's as popular as ever. Some prominent intellectuals have even expressed support for it, including the philosopher Daniel Dennett, the zoologist Lyall Watson and the naturalist Sir David Attenborough, who's hosted several aquatic-ape-boosting documentaries for the BBC.

What might be going on? Why does the theory endure?

Anthropologists lay a lot of the blame at the feet of Elaine Morgan, whom they've often accused of being like the charismatic leader of a pseudoscientific cult. If this is correct, Morgan's death in 2013 should mean that support for the theory will now slowly start fading. Only time will tell.

Although anthropologists don't like to admit this, the downfall of the savannah theory of human evolution in the 1990s did put a little bit of intellectual wind into the sails of the aquatic-ape model. It led several well respected theorists, including the evolutionary biologist Carsten Niemitz and the anthropologist Phillip Tobias, to suggest that it might be worth considering a middle ground between the dry and wet theories of our evolution. They were quite willing to jettison most of the aquatic-ape model, but they urged their colleagues to have a second look at wading as a cause of bipedalism, arguing that it was as plausible as any other explanation for why we walk upright. They didn't believe that our ancestors' exposure to water would necessarily have occurred along the coast. Early humans could have lived along riverbanks, at the edge of lakes, or in woodlands subject to seasonal flooding. But wherever it might have been, wading could have given them an adaptive advantage.

The strongest support for the theory, however, continues to be rooted in the issue of human uniqueness. It draws upon this very deep-seated intuition that we're different from other animals and particularly from other primates. There's just something odd about us, and this oddness, many feel, must have been the result of some extraordinary event in our evolution.

Anthropologists caution that this belief in our exceptionalism is misguided, that we're not as unique as we like to think we are. It's only our vanity making us believe that we are. Perhaps so, but sometimes it's hard not to be drawn in by the idea. You might particularly get this feeling if you ever spend the day at the beach, where you can witness our species splashing in the waves in all its Speedo-wearing glory. Faced with such a sight, the notion that we might really be the descendants of a bunch of soggy apes may not seem entirely unlikely after all.

Weird became true: the out-of-Africa theory

Humans live all around the world: in Africa, Asia, the Arctic, the jungles of South America and even remote islands in the Pacific. Our ability to adapt to survive in almost all of the Earth's environments is, in fact, one of our defining traits as a species. But where did we originate?

The accepted view among anthropologists, based upon a multitude of fossil evidence, is that our birthplace was Africa. The first of this evidence to emerge was the so-called Taung Child skull, discovered in South Africa in 1924 by Raymond Dart. It's regarded as one of the most important fossil finds of the twentieth century, but this wasn't always the case. When Dart initially presented it to the scientific community, their reaction was utterly dismissive. Leading British anthropologists waved it aside as inconsequential and rejected his argument that humans had originated in Africa. It took over a quarter of a century for them to change their minds.

Dart, as it turned out, hadn't actually wanted to be in South Africa. It was only by a quirk of fate that he happened to be there with the appropriate anatomical training to recognize the significance of the skull when it came to his attention. He was Australian by birth and had studied anatomy at Oxford. When he learned that the newly founded University of the Witwatersrand in Johannesburg needed someone to fill the position of Chair of Anatomy, he had no interest, fearing that going there would be like being sent into academic exile. He only applied for the job and accepted the position after some arm-twisting from his adviser.

Nevertheless, it turned out to be a great career move because it sent him directly on a path to make the fossil find of the century, which happened a mere two years after he arrived in South Africa. As far as anthropological finds go, it was probably one of the least strenuous ever, because it involved no searching or digging on his part. He never even had to leave his office. The skull was delivered directly to him as part of a group of fossils collected in a nearby mine called the Buxton Limeworks. He had earlier made arrangements with the mine to ship him anything interesting they found in the course of their work. When he received the box they eventually sent over, he opened it up and there was the skull, lying on the very top.

Dart immediately recognized its importance. The story goes that he was getting dressed to attend a friend's wedding when the box arrived, and his wife had to drag him away almost forcibly from it to get to the ceremony on time.

The skull was tiny, belonging to a child that must have lived hundreds of thousands of years ago, but its anatomy intriguingly combined features of both ape and human. Its brain was very small, like that of an ape, whereas its teeth were human-like. Dart could also see that its brain had human features, and he was able to determine that it must have walked upright, because faint markings revealed that its skull had been positioned upright on the spinal column, as opposed to hanging forward like that of an ape.

He believed that the creature, when it was alive, would have looked more ape-like than human, so he called it *Australopithecus*, meaning 'southern ape'. Its presence in Africa, he knew, was highly significant, because no such transitional fossils between apes and humans had yet been found there.

Dart promptly sent a report of his findings to the journal *Nature*, where it was published in 1925. But, to his dismay, instead of being hailed as a great find, it met with scathing criticism. Arthur

Keith, a distinguished British anthropologist, led the critics. He rejected the claim that the skull represented a transitional form between apes and humans as 'preposterous' – a word one doesn't use lightly in the science community when discussing a colleague's work. He declared that the skull was most likely that of a juvenile gorilla.

Part of this criticism was legitimate. The skeletons of primate infants tend to look somewhat similar. It's only with age that the differences between the species become distinct. So, the fact that it was the skull of a child made it difficult to identify precisely what it was. It was also a challenge to establish its geological age. The exact location of its discovery wasn't known, due to its delivery in a box, so Dart had to date it by anatomical clues. And Dart himself, at a mere thirty-two years old, was regarded as young and inexperienced.

A somewhat less legitimate reason for the criticism was that anthropologists simply didn't think humans had come from Africa. No less an authority than Charles Darwin himself had actually made the case for our African origin in the 1870s, noting that it made sense for humans to have originated where the other great apes lived, but by the 1920s this was no longer a popular belief. This was partly because Asia was then in vogue as the likely place of our origin, on account of early-human fossils that had recently been found there. Mostly, though, it was plain old racism. Elitist European scholars weren't inclined to acknowledge Africa as their ancestral home.

And then, as a final blow against the Taung Child, there was the matter of the Piltdown Man. This was an early-human fossil that had been found in Piltdown, England, in 1912, consisting of a skull that was large, like that of a modern-day human, but coupled with an ape-like jaw. The significance of this pairing was that it confirmed what many anthropologists had long suspected – that, in our evolutionary history, the first distinctively human feature to develop must have been a large brain. This made sense to them because they reasoned that intelligence is our most singular feature, so brain growth must have preceded all other anatomical changes.

The Taung Child, however, flew in the face of this reasoning. It combined a small brain with indications of bipedalism, suggesting our ancestors were walking upright long before they acquired large brains. In other words, the two fossils were telling different stories about our past. So, which to believe?

Given a choice between a big-brained British 'missing link' and a small-brained African one, British scholars came down firmly on the side of the 'first Englishman'. Keith was one of the strongest supporters of the importance of the Piltdown Man, which is why he so quickly concluded that the Taung Child was a gorilla, not an early human.

Dart grew depressed by the rejection of his find. His marriage fell apart, and for a while he gave up further anthropological research. He was lucky enough, however, to live to see his vindication, as new fossil evidence steadily accumulated that supported his interpretation of the Taung skull. These finds included the discovery of additional *Australopithecine* skeletons in Africa by Dart's colleague, Robert Broom.

By the late 1940s, the tide of intellectual opinion had swung decisively in Dart's favour. All the fossils being found were telling the same story of bipedalism first and brain development much later, and the oldest hominid fossils were consistently being found in Africa. It must have given Dart a sense of immense satisfaction when even his arch-nemesis Keith publicly recanted, writing in a 1947 letter to *Nature* that, 'I am now convinced . . . that Prof. Dart was right and that I was wrong.'

The final block of resistance crumbled away in the early 1950s with the stunning revelation that the Piltdown Man had been a crude forgery. A team of researchers at the British Museum had decided to examine it more closely, puzzled by how different it was from all the other fossil evidence. They discovered that its surface had been artificially stained – a clear sign of tampering. There's still enormous controversy about who the forger might have been. The leading candidate is Charles Dawson, the lawyer and amateur fossil hunter who had found it.

In hindsight, it should have been obvious that it was a fake. As early as 1915, the American scholar Gerrit Miller had noted that the mechanics of chewing made it impossible for the jaw and skull to belong together. British anthropologists had ignored his warning. The Piltdown fossil told them what they wanted to believe, so they disregarded the contrary evidence. Its unmasking as a fraud cleared the way for the full acceptance of the Taung skull and, by implication, the African origin of humanity.

What if we're descended from a pig–chimp hybrid?

In 1859, Charles Darwin published *On the Origin of Species* in which he challenged the old belief that species never changed by arguing that they actually evolved over time through a process of natural selection. Darwin realized this idea would shock people, and that it might particularly disturb religious sensibilities, so he carefully avoided delving into the implications of his theory for the origin of humans. He focused instead on species such as dogs and finches, and he saved the discussion of human evolution for twelve years later, in his book *The Descent of Man*.

Victorian readers, however, weren't fooled. They immediately put two and two together and figured out, as his critics put it, that Darwin was claiming man must be descended from the apes. This definitely riled up religious tensions. Did it ever! To this day, the descent of humans from apes remains a hot-button issue, with some people still simply refusing to believe it.

One can then only imagine the reaction if the hypothesis of the geneticist Eugene McCarthy were ever to gain mainstream scientific acceptance. He significantly cranks up the potential outrage factor, arguing that humans may not only be descended from apes, but from pigs as well. More specifically, he speculates that around six million years ago, a mating event may have occurred between a female chimpanzee and a male pig (or rather, between what would have been the ancestors of the present-day species), and that the resulting offspring was the progenitor of the human lineage.

McCarthy earned his Ph.D. in evolutionary genetics from the University of Georgia in 2003. Three years later, Oxford University Press published his *Handbook of Avian Hybrids of the World*, which was an encyclopaedic listing of thousands of reported crosses between different bird species, occurring in both wild and captive settings. It earned high marks from reviewers, and it suggested that McCarthy had a promising career ahead of him.

Which is to say that he initially seemed to be heading down a traditional path, and, if he had continued in that direction, he probably would by now be comfortably ensconced as a professor at some university, regularly churning out scholarly articles and enjoying the respect of his peers. But McCarthy veered sharply away from that course. He began entertaining strange and radical thoughts – the kind of speculations that eventually made him a pariah within academia.

There may have been a hint of the rebel already in his decision to specialize in the subject of hybridization. The term describes the phenomenon of two different species mating and producing an offspring. The most well-known example is the mule, which is the result of a cross between a male donkey and a female horse.

Hybridization is of great interest to plant breeders who want to produce new varieties of fruit and vegetables with potentially beneficial attributes. You might occasionally happen upon the results of such research in the produce aisle at the supermarket. Perhaps you've seen broccoflower (a hybrid of broccoli and cauliflower) or plumcots (plums plus apricots).

Crossing plants is one thing, though, but the idea of mixing animal species, particularly those more widely separated than donkeys and horses, is quite another. It causes all kinds of cultural red flags to go up. Imagine a cross of a dog and a cow, or of a goat and a horse, or of a human and anything else. Such mixtures evoke fascination and horror. The possibility of their existence seems monstrous and unnatural. They clash with our belief in the proper order of things.

Mythology is full of examples of disturbing hybrids. King Minos of Crete was said to have kept a Minotaur (a half-bull, half-human creature) in a labyrinth beneath his castle. A medieval legend speaks of the existence of a Vegetable Lamb of Tartary (a plant that grew sheep as its fruit). Perhaps the most famous hybrid of all is found in Christian tradition, with the Devil often depicted as half goat and half human.

These were the strange kinds of crosses that McCarthy began to contemplate. He wondered how far the phenomenon of hybridization could be pushed. How far apart could species be and still produce a viable offspring?

This led him to think about the role hybridization might play in evolution. The standard model is that evolution occurs through the steady accumulation of genetic mutations. Selective pressures in the environment then determine which mutations are preserved and which are not. When two populations of a species are geographically isolated from each other, this natural process causes them to evolve in gradually diverging ways until eventually they're no longer able to reproduce with each other. They become distinct species. The finches that Charles Darwin observed on the Galapagos Islands are a well-known example of this. Cut off from the mainland, they had undergone separate speciation.

McCarthy suspected that the reproductive barriers between species, even far-flung ones, weren't as rigid as almost all scientists assumed. If this was the case, then hybridization might occasionally jump-start evolution. A hybrid mating could suddenly introduce a whole new set of genes into a population, causing its development to take a dramatic new course.

If McCarthy had confined these evolutionary speculations to bird or plant species, they would have been controversial, but only among academics. Instead, he zeroed right in on the most inflammatory question possible: might hybridization have played a role in human evolution?

McCarthy didn't imagine every species to be a result of hybridization. He did, however, think it played a far greater role than most biologists credited it with, and he suspected that humans in particular might have been a product of the process. What led him to this conclusion was our low fertility rate in comparison with other mammals. Biological research, he pointed out, has found that, in a typical human male, up to 18.4 per cent of spermatozoa may be abnormally shaped and dysfunctional, compared to just 0.2 per cent in chimpanzees. Such a high rate of defective sperm has conventionally been attributed to clothing that affects scrotal temperature, but McCarthy noted that it's also a common trait of hybrids.

If humans are a hybrid, the next question was what two species might have parented us. McCarthy explained that, when biologists suspect a creature is a hybrid, there's a test they use to determine its parentage. First, they decide what species the creature is most like, and they assume that to have been one of the parents. Then, they make a list of all the characteristics that differentiate the suspected hybrid from its one known parent. With this list in hand, they try to match up the traits with another species. If they find a good fit, they figure they've probably found the other parent.

McCarthy went ahead and applied this test to *Homo sapiens*. We're clearly a lot like chimpanzees, so he assumed that species (the late Miocene era ancestor of it) was one of our parents. Then he combed through the scientific literature to produce a comprehensive list of ways that humans differ from chimps.

The resulting inventory included items such as our naked skin, the fact that we sweat copiously, our thick layer of subcutaneous fat, our protruding noses and our lightly pigmented eyes. Anatomical differences beneath the surface included our vocal cords and our kidneys. The latter have a uniquely shaped internal cavity, described as being 'multi-pyramidal' because it has numerous inward projections. There were also a variety of behavioural differences. Humans like swimming, snuggling and drinking alcohol, and chimps don't particularly care for any of these activities.

Having produced this list, McCarthy asked himself what creature displayed all these characteristics. He found only one candidate. It was *Sus scrofa*, the ordinary pig. In fact, the match was uncannily good . . .

A pig! McCarthy confessed that he himself was initially reluctant to take the idea seriously. It was only when he discovered that human and pig vocal cords actually look quite similar that he grew convinced there might be something to the crazy idea forming in his head. Once the thought took hold, he kept seeing more and more parallels between humans and pigs.

But wouldn't a pig–chimp hybrid be biologically impossible? Aren't there mechanisms on the cellular level that would prevent a viable offspring between such far-flung species? Wouldn't differing parental chromosome counts, for instance, prevent hybridization? And, even if an offspring did form, aren't hybrids usually sterile?

McCarthy cited these objections as common misconceptions about hybridization. He acknowledged that hybridization between far-flung species was improbable, but he insisted that it wasn't impossible. Nature might find a way to make it happen, and supposed reproductive barriers such as differing chromosome counts would do little to stop it. There are plenty of species with wildly differing numbers of chromosomes that manage to hybridize. Zebras, with forty-four chromosomes, and donkeys, with sixty-two, together produce baby 'zeedonks' just fine. What's more, hybrid offspring are often fertile. In his own work on bird hybrids, he had found that they were eight times more prevalent than sterile ones. It was true that a reduction in fertility was common, but this was exactly what had led him to suspect humans could be hybrids in the first place.

But what about the sheer unlikeness that a pig and a chimp would ever mate? McCarthy didn't believe this was an issue either, as long as a pig had been the father and a chimp the mother. Anatomically, only this combination would work, and behaviourally as well, because a male pig wouldn't have been fussy about what it mated with. Zoologists note that it's quite common for many animals to attempt to mate with what are euphemistically referred to as 'biologically inappropriate objects'. The female chimpanzee, McCarthy speculated, might have cowered submissively because she felt threatened. This was the romantic scenario he envisioned as the origin of the human species.

McCarthy imagined that this baby pig–chimp, once born, had been raised by chimps. When it reached maturity, it then mated with other chimps, as did its descendants. As the generations passed, this interbreeding caused the hybrid line to become progressively more chimp-like, which explained, McCarthy said, why humans today are far more primate than porcine. Most of the pig has been bred out of us. He argued that this same process would have concealed much of the genetic evidence of our pig ancestry, making it an extremely challenging task to detect it.

McCarthy self-published his pig–chimp hypothesis online in 2013. Realistically, that was the only way he was going to get it out in front of the public. No academic journal was going to touch the thing. Within weeks, the *Daily Mail* got wind of it and ran an article plastered with the headline, 'Humans evolved after a female chimpanzee mated with a pig'. This gained McCarthy a worldwide audience and instant notoriety.

Mainstream scientists went apoplectic. Several angrily suggested that, if McCarthy seriously believed pigs and primates could produce an offspring, he should try to impregnate a pig, preferably with his own sperm, and report back the results. McCarthy demurred.

His detractors had plenty of more specific criticisms. Many of them focused in on his list of ways that humans differed from chimps, blasting it as being cherry-picked to support his odd thesis. They noted that it omitted the two most obvious differences, which are that we walk on two legs and have big brains, neither of which are pig-like traits. The various similarities, they explained, while intriguing, were a result of convergent evolution, which is the phenomenon of unrelated species evolving similar features as a result of having faced similar selective pressures.

Critics also insisted that, despite what McCarthy claimed, there definitely were barriers to reproduction on a cellular level that would prevent pig–chimp hybridization. The pig and primate lineages split eighty million years ago. Too many differences had accumulated during that time to possibly be bridged. It was doubtful a pig sperm would even recognize a chimpanzee egg to be able to fertilize it. There was certainly no known example in animal biology of hybridization across such a vast taxonomic gap.

Then there was the lack of genetic evidence. Even if the descendants of the supposed pig–chimp hybrid had been interbreeding with chimps for many generations, as McCarthy contended, telltale clues of pig heritage should have remained in our DNA. Both the pig and human genomes had, by 2013, already been sequenced in their entirety, but analysis of the two had failed to reveal any obvious similarities. Pig sequences in our DNA were the kind of thing researchers would have noticed. As far as McCarthy's critics were concerned, this was the final and most decisive nail in the coffin of the pig–chimp hybrid.

The saga wasn't quite over, though. In 2015, there was a flurry of excitement among its fans (yes, the hypothesis boasts a fan base) when researchers revealed the discovery of an unexpected similarity between pig and human genetic elements called SINEs, or short interspersed elements. Did this mean there was genetic support for the pig–chimp hypothesis, after all? Was McCarthy about to be vindicated?

Not quite. McCarthy himself cautioned that the finding hardly proved his hypothesis, describing it as 'just one run in a nine-inning game'. Without far more substantial evidence, his hypothesis was doomed to remain relegated to the outer limits of the academic fringe.

But even if the theory is, as most scientists assume, wildly wrong, it does raise the provocative question of what the limits of hybridization actually are. How far apart *can* two species be and still produce an offspring?

You'll search biological literature in vain for a precise answer. There's a general consensus that eighty million years of taxonomic distance (as lies between pig and human) is far too big a gap to bridge, but what's the exact cut-off point? Forty million years? Four million? Two million? It seems to depend on what you're trying to match.

Lions and tigers, whose lineages split around four million years ago, manage to produce offspring. Most biologists don't think that humans and chimpanzees, however, who diverged six million years ago, could hybridize. Though, really, there hasn't been much effort to test this assumption. But it was confirmed in 2012 that our ancestors did interbreed with Neanderthals, from whom we diverged about 500,000 years ago, which makes many of us Neanderthal–human hybrids.

This data suggests that only relatively recently separated species can hybridize, but there are outliers that complicate this picture, such as guinea fowl and chickens, which can produce fertile baby guinhens even though their lineages diverged fifty-four million years ago. If you look at hybridization among plants, all bets are off.

And, with genetic engineering, almost anything may now be possible. Scientists are hybridizing species that would never come across each other in nature. As it turns out, one species that researchers are particularly interested in is the pig. Many pig organs genuinely are similar to human organs, as McCarthy pointed out. This has led to great interest in the possibility of transplanting pig organs into people. If this were possible, it could solve the organ-shortage crisis.

Serious problems stand in the way of this, not the least of which is how to stop human immune systems from rejecting pig organs. One possible solution, though, which has already had billions of dollars devoted to it, is to breed pigs that are more human, on a cellular level. Researchers at the Salk Institute and the University of California are actively working to achieve this and they've already succeeded at creating specimens of human–pig chimeras.

Which is to say that, even if a pig–chimp hybrid didn't come into existence six million years ago, unusual creatures that are mixtures of the two do now exist in laboratories. They just happen to be the creation of science, not nature.

What if hallucinogenic drugs made us human?

When ingested, the mushroom *Psilocybe cubensis* begins to cause noticeable effects after about twenty minutes. These can vary greatly from person to person, but it's common for the physical ones to include pupil dilation and an increased heart rate. The psychological ones may involve dizziness, confusion, visual hallucinations, distortions of space and time and a profound feeling of being at one with the cosmos.

These effects are only temporary. For most, they wear off after four to six hours, but they can leave a powerful lasting impression. The author and ethnobotanist Terence McKenna was particularly moved by his experiences. So much so that he began to suspect *Psilocybe cubensis* might have played a more far-reaching role in the history of our species than anyone had previously suspected. What if, he wondered, the mushroom was the reason for the emergence of human intelligence?

McKenna detailed this speculation in his 1992 book *Food of the Gods*. He envisioned our distant ancestors munching on mind-altering mushrooms and having their brains literally expanded, over successive generations, by the subsequent psychedelic experience. He called this his stoned-ape theory of human evolution.

The central mystery touched upon by the stoned-ape theory is the remarkable evolutionary development of the human brain. Two million years ago, the brains of our hominid ancestors were only a third of the size of the present-day average and were just a little larger than the brains of modern chimpanzees. Then they started growing, rapidly. As far as we know, it's the only time in the history of evolution that a species has experienced such rapid brain growth, and the end result was that humans gained brains that were larger, relative to body size, than those of any other creature on Earth. What could have caused this extraordinary growth?

Palaeoanthropologists have proposed a number of possibilities, such as tool use, language, communal hunting and even our sociality as a species. The problem is that it's been impossible to come up with a definitive answer because there's so little evidence to go on. Brains don't fossilize, and skulls, which do fossilize, can only reveal so much about the grey matter they housed. Given this meagre amount of material to work with, consensus has eluded researchers. Which is why a crack has remained open for a more unorthodox possibility, and along came McKenna's stoned-ape theory.

Although McKenna is often described as an ethnobotanist, which sounds scientific, he didn't have any formal training. He was a self-taught visionary and intellectual. After a conventional childhood in a small town in Colorado, McKenna headed to the University of California, Berkeley, in the mid-1960s and fell in with the counterculture. He then took off to travel the world, eventually ending up in the Amazon jungle, where he first experienced psychedelic mushrooms. It changed his life. He returned home to the United States and, in 1976, co-authored *Psilocybin: Magic Mushroom Grower's Guide* with his brother Dennis. It sold over 100,000 copies, and he subsequently developed a career as a lecturer and writer, making it his mission in life to evangelize on behalf of psychedelic drugs. His stoned-ape theory was, in a way, the culmination of this effort. It was his attempt to offer a naturalistic, scientific case for how magic mushrooms could have supercharged the evolution of the human brain.

*

According to McKenna, humanity's psychedelic brain evolution began several million years ago (he was vague on exact dates) when proto-humans wandered out of the steaming jungles of Africa onto the dry grasslands. Our ancestors, at this stage, as he tells it, were rather low achievers. They scavenged for food wherever they could find it, often following the herds of wild cattle that migrated across the savannah.

Then, one fateful day, one of these proto-humans made a serendipitous discovery. As he trailed behind a herd, weaving in between piles of manure, he spotted a mushroom growing in a dung heap. He reached down, plucked out the fungi and popped it in his mouth. The result was an experience of startling novelty, because this wasn't just any old mushroom. It was a magic mushroom, *Psilocybe cubensis*, containing the potent hallucinogen psilocybin. This first accidental psychonaut had discovered the 'visionary fungi of the African grasslands'. Soon, all his companions were seeking out these dung-growing mushrooms, transforming themselves into the stoned apes of the theory.

McKenna believed that these mushrooms didn't merely excite our ancestors with pleasurable sensations, they also provided them with adaptive evolutionary advantages. He referred to studies by the psychiatrist Roland Fischer suggesting that, at low levels, the mushrooms improved visual acuity, particularly edge detection, which aided hunting. They served as chemical binoculars. At slightly higher levels, the fungi increased sexual arousal, encouraging those who ate them to mate more often and produce more offspring than abstainers. At these quantities, the mushrooms also mellowed the fierce individuality of the male hunters, calming them down and helping to promote communal caring for the young.

At even higher levels, the mushrooms yielded 'full-blown shamanic ecstasy'. Here, the brain development really came into play. One of the known effects of psilocybin, noted McKenna, is that it causes the senses to overlap and blend in peculiar ways. It seems to reorganize the brain's information-processing capabilities.

McKenna argued that this perceptual rewiring might have broken down mental barriers, catalysing the development of imagination, self-reflection, symbolic thought and, perhaps most importantly, language. And, if it did so, if it encouraged early humans to vocalize and interpret the sounds coming out of their mouths in new ways, it could, over many millennia, have led to an increase in brain size. As McKenna put it, we may 'literally have eaten our way to higher consciousness.'

In McKenna's chronology, the period of active human–mushroom symbiosis lasted for almost two million years, from the era of our distant ancestor *Homo habilis* right up to the dawn of civilization. He deemed this to have been a golden age in our history, the era of what he called a 'partnership society', when humans developed their full brainpower, nourished by the wisdom of the fungi. This phase ended 12,000 years ago, when the mushrooms grew scarce because of climate change, and our ancestors settled down to adopt agriculture.

However, they missed the mushrooms, and sought other drugs to take their place. What they found was alcohol. For McKenna, this represented a tragic fall from grace, our banishment from the Garden of Eden, because alcohol, of which he was no fan, promoted aggression and hierarchy, giving rise to a 'dominator culture' that has ruled for the past twelve millennia.

McKenna's history of our species ultimately turned into a tale of lost innocence, leading to the present day, in which, by his diagnosis, we live alienated from nature and from one another. The cure he prescribed was to embrace once more the shamanic wisdom of the mushroom, thereby reuniting with nature and regaining the paradise of the partnership society we lost.

Major publications reviewed *Food of the Gods*, including the *Los Angeles Times* and the *Washington Post*, as well as scientific journals such as *Nature* and *American Scientist*. McKenna couldn't claim that he was ignored. The good news for him was that many critics praised his gift for language. The bad news was that almost everyone panned his scientific claims. They just didn't buy his thesis that drugs had made us human.

A recurring criticism was that he only marginally engaged with the existing scientific literature. He romped through various disciplines – anthropology, archaeology, psychology, mycology – using whatever facts suited his purpose, but he didn't delve deeply into any one subject area. The book focused on sweeping ideas rather than the minutiae of scholarship.

Another complaint was that his social and political views interfered with his science. He made no attempt at all to be scientifically impartial. Instead, he openly lobbied for liberalizing the drug policies of the industrialized world, insisting that psychoactive drugs could play a positive role in society and shouldn't be criminalized. The result, complained critics, was that his book ended up reading more like pro-drug propaganda than science.

More seriously, critics accused him of misrepresenting research. For example, the study by Roland Fischer that he referenced, claiming it had shown that psilocybin improved visual acuity, actually said no such thing. It had shown that psilocybin altered vision, but it made no suggestion of an improvement. The 'chemical binoculars' that McKenna imagined aiding Palaeolithic hunters didn't actually exist.

Then there was the sheer outrageousness of his argument. The entire premise seemed totally outlandish and, frankly, silly. The scientific community, therefore, reached its verdict. No one could deny that McKenna had a way with words and a gift for coming up with provocative ideas, but ultimately it was deemed that he lacked the scholarly rigour to produce a compelling, credible argument. The charitable interpretation was that the stoned-ape theory was an entertaining fable dreamed up by someone overenthusiastic about the positive value of psychedelics. The less charitable interpretation was that it was pseudoscientific foolishness.

So, McKenna definitely failed to sell his theory to mainstream science, but was this a case where the overenthusiasm and inexperience of the messenger got in the way of the message? Does the stoned-ape theory deserve to be taken seriously despite its rejection? A handful of researchers think so.

These supporters tend to come from the field of psychedelic-drug research, which means they're more inclined than most to attach great significance to psychedelics. Nevertheless, they're convinced that hallucinogenic mushrooms could have played a role in human evolution and, since McKenna died in 2000, they've tried to keep the stoned-ape theory alive in his absence.

One of these fans is the mycologist Paul Stamets, considered to be a leading authority on mushrooms and psychedelics. In April 2017, at an academic conference in California, he declared his belief that the stoned-ape theory was 'right on', to which the audience responded with enthusiastic applause. Another fan is Terence McKenna's brother, Dennis – hardly an unbiased source, though unlike Terence he has full-blown scientific credentials, with a doctorate in botanical sciences from the University of British Columbia.

Their argument for taking the theory seriously partially focuses on the growing scientific appreciation of the power of psychedelics. New research continues to reveal the dramatic effect these drugs have on the brain. Recent studies using functional magnetic resonance imaging (fMRI) have shown that they stimulate deep connections between parts of the brain that wouldn't normally be in communication, and the effects seem to be extremely long-lasting. Many users report that the experience of taking them is permanently life-changing.

There's also the fact that these powerful psychoactive substances were certainly present in the environment inhabited by our ancestors. *Psilocybe cubensis* is indigenous to the tropics and subtropics, where it grows in the dung of a variety of species, including elephants, zebra, antelopes, buffalo and cows. A curious proto-human would have needed only to pick a mushroom up and pop it in his mouth to experience its effect.

To the stoned-ape advocates it seems only natural to connect the dots between these two things and conclude that hallucinogens could plausibly have played a role in the human brain's sudden development.

The objection from critics is that, no matter how powerful and readily available these drugs might have been, there's no apparent reason why they would have influenced human evolution. The theory's advocates respond by returning to McKenna's argument that the drugs might have aided the acquisition of language.

Consider what differentiates animal communication from that of humans. Many animals have simple forms of communication. Dogs bark, and vervet monkeys have been observed making warning calls specific to different predators. But, in these cases, one sound always has one specific meaning. Humans, however, have mastered complex, symbolic forms of language that allow us to communicate highly abstract messages. We mix sounds and concepts together to create an endless variety of meanings. How did our ancestors learn this trick?

Dennis McKenna points out, as his brother did, that the signature effect of psilocybin is synaesthesia. It causes a mixing of the senses. Parts of our brain connect that aren't normally in tune with each other. This seems like precisely the kind of nudge that might have been needed to help our hominid ancestors make the leap from simple sounds to more complex, symbolic language. It's tempting to imagine one of them ingesting *Psilocybe cubensis* and then having a dim awareness form in his head that those sounds emerging from his mouth could have different meanings if combined in new ways. Dennis McKenna has no doubt of the link. He declares it as a fact: '[Psilocybin] taught us language. It taught us how to think.'

Along similar lines, author Dorion Sagan, son of Carl Sagan, has drawn attention to the curious similarities between language and psychedelic drugs. Hallucinogens create images in a person's head, but so does language. We use words to form pictures in each other's brains. Poetry and song can produce a flood of powerful visual imagery. In this sense, Sagan suggested, language is actually a form of 'consensual hallucination'. It's the ultimate hallucino-genic drug.

Perhaps these are just coincidental similarities. We'll probably never know for sure. It's hard to imagine a palaeontological find-ing or psychological experiment that could settle the debate one way or another. But it is intriguing to think that there might be a deep underlying connection between hallucinogens and language.

And there's one final concept that persuades some the stoned-ape theory might be worth a second look. It's the idea that the human mind is such an extraordinary thing that to account for it perhaps we need to look beyond standard evolutionary explan-ations and consider the possibility that a remarkable chance event brought it into being.

Terence McKenna often spoke of mushrooms and early humans forming a symbiotic relationship. It's easy to dismiss this as one of his poetic flourishes, but it does recall the anthropological con-cept of efflorescence, a term used to describe how contact between two cultures will often lead to a blossoming of creativity that pro-duces unexpected results. One example is how global trade spread Chinese innovations, such as gunpowder and printing, to Europe, where they were developed in dramatically new ways.

McKenna's theory asks us to consider that something similar might have occurred in Africa two million years ago. Imagine that two species from different biological kingdoms fortuitously crossed paths. The complex chemical system of a fungus might have encountered a primate that was uniquely ready to benefit from it. The result was an unexpected moment of evolutionary efflorescence. New mental pathways and dormant abilities stirred in the primitive ape brain, and then slowly our ancestors began to hallucinate their way to higher consciousness.

Weird became true: cave art

To step inside the Cave of Altamira, located on the northern coast of Spain, is like walking into a prehistoric cathedral. Its ceiling and walls are covered with stunning depictions of bison, aurochs, deer and horses, all drawn well over 10,000 years ago by artists whose identities have been lost in the mists of time. These magnificent Palaeolithic paintings are today considered to be among the artistic wonders of the world, but, remarkably, they weren't always held in such high esteem. When the cave was discovered in the late nineteenth century, it was the first time such elaborate prehistoric art had ever been seen, and the leading experts in the young field of prehistory were singularly unimpressed. They promptly declared that it was impossible for cavemen to have created artwork of this kind and dismissed it as a modern-day hoax. It took over twenty years before they finally came around to acknowledging its authenticity and significance.

The paintings were discovered by a wealthy Spanish landowner, Don Marcelino Sanz de Sautuola. In 1878, he had attended the Paris World Exposition, where he saw a collection of prehistoric artefacts found in France. His curiosity stirred as he remembered that, several years before, in 1868, a hunter searching for his dog had found the opening of a cave on his estate. Sautuola wondered if this cave might contain any prehistoric artefacts.

It wasn't until the following year, though, that he got around to following up on this thought and looked more closely at the cave. When he dug near its entrance, he quickly found some flint tools and bones, which excited him. The following day, he brought along his nine-year-old daughter, Maria, so that she could play as he worked. He bent down to resume his digging, and Maria ran into the cave straight away to explore. A few minutes later, Sautuola heard her shout out, 'Look, Papa, bison!'

He had gone into the cave earlier by himself, but had been so intent on what he might find in the ground that it had never occurred to him to look up at its ceiling. As a result, he had missed the almost life-size multicoloured bison drawn there, until Maria's call drew his attention to them.

This charming tale of a child's role in such an important discovery has now become one of the most popular stories in modern archaeology. It's often told as a reminder not to be too focused on the immediate task at hand. Remember to stop and consider the bigger picture every now and then!

Once he had seen the paintings, Sautuola instantly realized their significance, and he diligently set to work to spread the word of their discovery. He prepared a pamphlet in which he carefully described them, arguing that they had to be prehistoric. He also gained the assistance of Juan Vilanova y Piera, a professor at the University of Madrid, who was similarly impressed by them. Together, the two men travelled to a series of academic conferences in Spain, Germany and France in order to present information about the cave. They had anticipated enormous interest in the find. Instead, they were greeted with open scepticism and contempt.

It didn't help that Sautuola was an unknown amateur. If an experienced academic with a well-established reputation had made the discovery, perhaps the reception would have been different. Equally, nothing on the scale of this cave art had been seen before. It seemed almost too extraordinary to be true, and, since the discipline of prehistory was at the time only a few decades old, its practitioners were paranoid of being discredited by falling for a hoax.

Most of all, though, the cave art contradicted the prevailing image of our Stone Age ancestors. This image had been shaped by Darwin's recently published theory of evolution, which argued that humans had descended from apes. Scholars of prehistory took it as a matter of faith, therefore, that our distant, cave-dwelling ancestors must have been closer in behaviour to chimpanzees than to humans. The Altamira cave art, however, was clearly not the work of beings anything like chimps. It had been created by skilled artists, every bit the equal of modern painters.

Two influential French academics, Émile Cartailhac and Gabriel de Mortillet, led the opposition to the discovery. Without having set foot in the cave, and even refusing to do so, Cartailhac declared its paintings to be 'a vulgar joke by a hack artist'. Both academics denounced the cave art as a hoax deviously conceived as an attack on evolutionary theory. Cartailhac pinned the blame on conservative Spanish clerics, while Mortillet suspected it to be the work of antievolutionist Spanish Jesuits.

An element of nationalistic jealousy may also have influenced Cartailhac and Mortillet. They simply didn't want to credit ancient Spaniards with having created such magnificent work. If the cave had been located in France, their opinion of it might have been more charitable.

To support their accusation that it was all a fraud, these scholars pointed out unusual features of the cave and the paintings, such as the lack of smoke marks on the ceiling. How, they asked, could cavemen have painted such works without burning a fire to see? This proved, they declared, that the paintings were of recent origin. The rest of the academic community obediently fell in line behind their lead.

Sautuola found his reputation to be in tatters. He protested that animal fat, when burned, doesn't produce much smoke, but he was ignored. Instead of being hailed as a great discoverer, he was cast aside as a charlatan. He was even banned from attending further academic conferences.

It was only the discovery of more cave art that eventually

vindicated him. In 1895, engravings and paintings similar to those at Altamira – including depictions of bears, mammoths, lions and more bison – were found in caves at Les Combarelles and Font-de-Gaume, in France. This time, no one doubted the artwork was Palaeolithic, and its location in France apparently placated the French scholars. With their nationalistic pride calmed, they reconsidered the authenticity of the Altamira paintings.

It wasn't until 1902, though, that full scholarly recognition came. In that year, Cartailhac penned a public apology to Sautuola, titled, 'Mea culpa d'un sceptique' ('Mea culpa of a sceptic'). Unfortunately, it was too late for the wronged discoverer to receive it; he had died in 1888. Cartailhac did, however, travel to Spain to apologize in person to Sautuola's daughter Maria, now an adult. While he was there, he set foot for the first time in the Cave of Altamira and finally gazed upon the amazing paintings, the authenticity of which he had, for so long, refused to acknowledge.

What if humanity is getting dumber?

What would happen if we were able to pluck a caveman out of the Stone Age and drop him down into the twenty-first century? Would we find that he was smarter or dumber than us?

Cavemen don't get a lot of intellectual respect. The stereotypical image is that they were brutish, stupid, club-wielding oafs. Most people might assume that a time-transported caveman would be out of his intellectual league here in the present. Several scientists, however, have put forward the disturbing suggestion that the caveman would actually mentally outshine modern humans by a wide margin. His mind would be sharp, clear-sighted, able to grasp complex ideas easily, emotionally stable and gifted with a powerful memory. The reason for his mental superiority, according to these researchers, is that the brainpower of *Homo sapiens* has been heading downhill for the past 12,000 years. The late Stone Age, they say, marked the intellectual high-water mark of our species. This is known as the idiocracy theory, after the 2006 movie of that name about a future world in which humanity has grown morbidly stupid.

The first clue that we may be getting dumber emerged from the field of palaeoanthropology, the branch of archaeology that focuses on the evolutionary history of humans. By the end of the twentieth century, researchers had gathered enough data about the size of the human brain during the course of our evolution to reveal a disturbing fact. Over the three-million-year span of the Stone Age, the average brain size of our ancestors grew substantially, but in the late Stone Age (50,000 to 12,000 years ago) it peaked. Since then, it's been getting smaller, by quite a bit.

Twenty thousand years ago, for example, the Cro-Magnon people living in Europe boasted an average brain size of 1,500 cubic centimetres. Today, brain size is about 1,350 cubic centimetres, which represents a 10 per cent decline. In visual terms, we've lost a chunk of grey matter about the size of a tennis ball. While 20,000 years may sound like a long time, on the timescale of evolution it's a blink of an eye. Which means that, not only has the human brain been shrinking, but it's been doing so rapidly.

These gloomy statistics about brain size aren't being disputed. The real question is, what does the shrinkage mean? Does it indicate that we're getting dumber? Is it as simple as that?

Maybe not. Anatomists are quick to point out that, while there's a loose correlation between brain size and intelligence, there's not a strict one. The brain of a genius, for example, isn't necessarily going to be larger than the brain of a dullard. Albert Einstein had an average-sized brain.

When comparing brains between species, the most important factor is the ratio of brain volume to body mass. This is known as the encephalization quotient, or EQ. The more brain per ounce of body weight a species has, the smarter it generally is. A small creature with a big brain will, therefore, probably be smarter than a big creature with a big brain. Humans have more brain per ounce than any other species in the animal kingdom, and this seems to be why we're the Earth's smartest species. Or, at least, its most dangerous.

So, perhaps the brains of the Cro-Magnons were larger simply because their bodies were also larger. That would be one explanation for our shrinking brains. We could chalk up our small craniums to the general downsizing of our bodies as a whole.

Unfortunately, no dice. Recent studies indicate that Cro-Magnon brains were proportionally larger than ours, even when we consider their larger body size. The human brain has shrunk far more than the human body has. EQ doesn't get us off the hook.

This finding led David Geary, a cognitive scientist at the University of Missouri, to conclude that we really may be getting stupider. In his 2005 book, *The Origin of Mind*, he raised the possibility that intellectual abilities may have declined across the human population as a whole. He was the first to jokingly refer to this as the idiocracy theory.

Geary suspects that our cranial downsizing was related to the adoption of agriculture in the late Neolithic period, around 12,000 years ago. He charted human brain size over time and discovered that it was at this exact historical moment, as population densities increased, that the human brain began noticeably shrinking. This trend repeated throughout the world, wherever complex societies emerged.

Geary reasons that, cavemen stereotypes aside, it took high intelligence to survive in the Stone Age, when our ancestors lived as hunter-gatherers. There was a broad array of skills that had to be mastered, such as how to find and identify edible food consistently, how to avoid predators and how to hunt well. You couldn't be a dummy and stay alive. If you made a mistake, you didn't get a slap on the wrist. You died. This placed enormous selective pressure on intelligence.

When our species adopted agriculture, this provided a more reliable supply of food, and populations began to swell. Towns formed around the fields, and then cities took their place. Civilizations were born. Often, even the most positive developments, however, can have unintended consequences. In this case, agriculture eased the pressure on our species to be smart. It created a kind of safety net, so the less quick-witted remained alive and passed on their genes. Over the span of millennia, this absence of selective pressure steadily lowered the average brain size of our species. The end result has been that we're getting dumber.

*

As if having a shrinking brain wasn't bad enough, another argument for the idiocracy theory has emerged from the discipline of genetics. In a two-part article that appeared in the January 2013 issue of *Trends in Genetics*, Stanford professor Gerald Crabtree argued that a straightforward genetic analysis also suggests we've been getting dumber since the Stone Age. He called it the 'fragile intellect' phenomenon.

The problem, as he described it, is that our intelligence requires many genes to be functioning just right. He estimated that between 2,000 and 5,000 genes are involved in maintaining our intellectual abilities, representing 10 per cent of our entire genome. According to Crabtree, the sheer number of genes associated with intelligence presents a problem, because it allows a greater chance for harmful mutations to accumulate. This makes it a fragile trait, prone to fading away over the span of generations.

To understand this argument, think of the game of Telephone (or Chinese Whispers, as it's sometimes called), in which a message is transmitted along a line of people, each individual whispering the message into the ear of their neighbour. If the message is simple – one word, perhaps – there's a good chance it will arrive at the end of the line unchanged. The longer the message, however, the higher the chance for alteration. The analogy isn't exact, but the same basic concept holds with genes. The more genes that are involved with a trait, the greater the chance for mutations to accumulate as the genetic information is transferred from one generation to the next, and if nothing is done to weed out these mutations, their numbers will multiply, eventually causing a negative impact on the trait – in this case, intelligence.

If this is true, how did intelligence ever evolve in the first place? The answer, explained Crabtree, was because of the force of selective pressure during the hundreds of thousands of years that our ancestors lived as hunter-gatherers. As Geary had previously noted, there was extreme selective pressure on intelligence, because only those with the wits to survive in a harsh, unforgiving environment lived to pass their genes on. In these circumstances, humans grew very smart, but they paid a terrible price with their lives.

Crabtree, like Geary, sees the adoption of agriculture as a pivotal moment in our evolutionary history. It shielded our species from the intense selective pressure required to maintain the genes for intelligence intact. We flourished, but the cost of this success has been a slow inevitable decline in our intellect. To put it bluntly, dumb people are no longer being culled from the human herd. Crabtree estimated that, compared to a late-Stone Age hunter-gatherer, each of us has between two and six deleterious mutations in our intelligence-related genes, which doesn't sound like a lot, but it may be enough to make us generally more slow-witted than our ancestors. These mutations will just keep accumulating in future generations, so our descendants can look forward to being even more intellectually dull than we are.

So, both palaeoanthropology and genetics suggest we may be getting dumber. 'But wait a second!' you may be tempted to say. How is it possible for anyone to seriously argue that we're getting dumber? Just look at all the amazing things our species has done in the past hundred years. We sent men to the moon, cracked the genetic code and created computers. If anything, humans seem to be getting more intelligent, not less.

This is a common objection to the idiocracy theory. How can we be getting dumber if, all around, people are doing increasingly smart things? The answer to this seeming paradox comes from the humanities, and it forms the final leg of the idiocracy theory. It's called the theory of collective learning. The researchers who formulated it, including the historian David Christian, never intended it to be used to support a theory of our declining intellect, but it could provide an explanation.

Collective learning describes the ability of humans to accumulate and share information as a group, rather than just individually. As soon as one person learns something, everyone in the group can learn and benefit from that information.

Christian has argued that collective learning is the defining feature of our species, since humans alone have mastered this trick. Early humans acquired the ability when they first invented symbolic language, tens of thousands of years ago. This allowed them to share complex, abstract ideas with each other, and then to transmit that knowledge to future generations. In this way, as a species, we began to accumulate more and more information, and the more we had, the easier it became to acquire still more. The process took on a momentum of its own.

Soon, we invented technologies that amplified our ability to share and accumulate information. We were, in effect, able to outsource some of the functions of our brain, such as memory. Writing was the most powerful of these technologies, followed by printing and now computers. Thanks to these technological innovations, we're amassing information at an almost exponentially increasing rate.

The thing about collective learning, however, is that it's a group phenomenon. As individuals, we may or may not be smart, but when we network together we become very intelligent as a collective entity. In other words, we shouldn't look at all the advanced technology invented in the past century and conclude that we must be the smartest humans ever to have lived. Our achievements are only possible because we're the beneficiaries of knowledge amassed through generations of collective learning.

In fact, we very well could be getting dumber as individuals, and yet the force of collective learning would continue to drive our civilization onward to more innovation and complexity.

The idiocracy theory has a certain gloomy logic to it. If you're feeling pessimistic about the current state of the world, you may even feel it has the self-evident ring of truth. But, if you're more optimistic about the state of humanity, then rest assured that mainstream science doesn't put much credence in the notion that our mental powers are on the wane.

There are, for instance, other plausible explanations for why our brains have got smaller since the Stone Age. Harvard primatologist Richard Wrangham argues that it may simply be a symptom of self-domestication. Animal researchers have discovered that domesticated breeds always have smaller brains than non-domesticated or wild breeds. Dogs, for example, have smaller brains than wolves.

Researchers believe that the link between domestication and small brains comes about because domestication selects for less aggressive individuals. Breeders favour individuals that are friendly and easy to get along with, and, as it turns out, being cooperative is a juvenile trait associated with young brains. Aggression in wild species emerges with adulthood. By selecting for friendliness, therefore, breeders are inadvertently selecting for individuals who retain a juvenile, smaller brain as adults.

When applied to human brain size, the argument goes that, as population density increased, it became more important for people to get along with each other. Overt aggression undermined the stability of large groups, and so the most combative individuals were systematically eliminated, often by execution. In effect, the human species domesticated itself, and as a result our brains got smaller. This doesn't mean, though, that we got more stupid.

As for Crabtree's 'fragile intellect' argument, his fellow geneticists panned it. The general theme of the counterargument was to deny that the transition to agriculture had relaxed selective pressure for intelligence. Kevin Mitchell of Trinity College Dublin argued that higher intelligence is associated with a lower risk of death from a wide range of causes, including cardiovascular disease, suicide, homicide and accidents. So, smarter individuals continue to enjoy greater reproductive success. Furthermore, the complexity of social interactions in modern society may place a higher selective pressure on intelligence because it serves as an indicator of general fitness.

A group of researchers from the Max Planck Institute of Molecular Cell Biology and Genetics echoed these criticisms, adding that intelligence isn't a fragile trait, as Crabtree feared. In fact, it seems to be quite robust, genetically speaking – the reason being that the large number of genes involved in intelligence aren't devoted solely to that trait. Instead, many of them are also associated with other vital functions, such as cell division and glucose transport. Strong selective pressure continues to preserve them.

A larger reason lurks behind the scientific distaste for the idiocracy theory. Many fear it raises the disturbing spectre of eugenics – the idea that scientific or political committees should decide who gets to breed in order to ensure that only the 'best' people pass on their genes. There was a time in the nineteenth and early twentieth century when many leading scientists were advocates of eugenics. It was a dark time in the history of science, and no one wants to revisit it.

Crabtree, for his part, insisted he was in no way an advocate of this. He raised the issue of our possibly fragile intellect, he said, not to make a case for social change, but simply as a matter of academic curiosity. In fact, he noted that, with our current state of knowledge, there's nothing obvious we could do about the problem if it does exist. We simply have to accept the situation.

Or do we? We may already be doing something to reverse the trend. It turns out that throughout the history of our species, one of the great limiting factors on the size of the human head, and therefore of the brain, has been the size of women's pelvises. Very large-headed babies couldn't fit through their mother's pelvis during birth. In the past, this meant that large-headed babies would often die during childbirth, but now, thanks to the ability of doctors to safely perform Caesarean sections, they no longer face that risk. We've removed the ancient limit on head size, and researchers suspect that this has already had an impact on our evolution. In the past 150 years, there's been a measurable increase in the average head size.

This means that, if the human head can now grow as big as it wants, a larger brain size might follow. Just by dumb luck, we may have saved ourselves from idiocracy.

CHAPTER FIVE

Mushroom Gods and Phantom Time

During the past two million years, human species of all shapes and sizes could have been encountered throughout the world. For most of this time, the long-legged *Homo erectus* roamed in Eurasia, and the brawny Neanderthals, more recently, occupied the same region. Further away, on the Indonesian island of Flores, the small, hobbit-like *Homo floresiensis* made their home.

None of these species still survive. It's not clear what became of them. All we know for sure is that only one human species remains: ourselves, *Homo sapiens*. Our ancestors first emerged from Africa around 100,000 years ago – at which time, *Homo erectus* had already vanished, although both the Neanderthals and *Homo floresiensis* were still around – and they rapidly migrated around the world.

A distinct cultural change became apparent among our ancestors, some 50,000 years ago. No longer did they just make stone axes and spears. Suddenly, they began carving intricate tools out of bone, covering the interiors of caves with magnificent paintings, and even engaging in long-distance trade. Still, they lived a nomadic life, hunting and foraging for food. Then, around 12,000 years ago, some of them decided to start planting crops. Soon, they had settled down to live besides these crops, and they became farmers. This marked a pivotal moment in our history because it eventually led to the rise of civilization, which is the era we'll examine in this final section.

The term 'civilization' is related to the Latin word *civitas*, meaning 'city', the first of which appeared around 5,500 years ago, developing out of the earlier agrarian villages. The first forms of writing emerged in Mesopotamia and Egypt just a few hundred years after the rise of cities. It was a technology that allowed people to keep track of the resources they were accumulating, as well as to record their beliefs and memorialize important events.

As a result of the existence of these written records, the researchers who study this era – social scientists such as archaeologists, anthropologists, philologists, historians and psychologists – don't have to rely solely on teasing out clues from fossils and other physical remains to figure out what happened. They can read first-hand accounts. Unfortunately, this doesn't translate into greater certainty. Humans are notoriously unreliable as sources of evidence – we lie, exaggerate, embellish, misremember and misinterpret events – so disagreements abound about the historical record, and strange theories proliferate, dreamed up by those who are convinced that, to get at the real truth of this era, we need to read deeply between the lines.

What if ancient humans were directed by hallucinations?

Imagine that engineers could build a society populated by sophisticated biological robots. From the outside, it would look exactly like our own. You would see people driving to work, doing their jobs, dining at restaurants, going home at night and falling asleep, but if you ever stopped one of these robots and asked it to explain why it had decided to do what it was doing, it would have no answer. It never decided to do anything. There was no conscious thought involved. It had simply been following what it had been programmed to do.

In 1976, the bicameral-mind theory of Princeton psychologist Julian Jaynes introduced the idea that this is very much like the way humans functioned until quite recently in our history – about 3,000 years ago. Jaynes didn't think our ancestors were robots, of course, but he did argue they had no self-awareness or capacity for introspection. They built cities, farmed fields and fought wars, but they did so without conscious planning. They acted like automatons. If asked why they behaved as they did, they wouldn't have been able to answer. So, how did they make decisions? This was the strangest part of Jaynes's theory. He maintained that they were guided by voices in their heads – auditory hallucinations – that told them what to do. They obediently followed these instructions, as they believed these were the voices of the gods.

*

As a young researcher, Jaynes became interested in the mystery of human consciousness. It wasn't consciousness in terms of being awake or aware of our surroundings that intrigued him. Instead, he was fascinated by the consciousness that forms the decision-making part of our brain. This could be described as our introspective self-awareness, or the train of thought that runs through our minds while we're awake, dwelling upon things we did in the past, replaying scenes in our mind and anticipating events in the future.

This kind of consciousness seems to be a uniquely human phenomenon. It's not possible to know exactly how animals think, but they seem to live in the moment, relying on more instinctual behaviours to make decisions, whereas we have a layer of self-awareness that floats on top of our instincts. Jaynes wondered where this came from.

To explore this question, he initially followed what was then the traditional research method. He studied animal behaviour, conducting maze experiments and other psychological tests on worms, reptiles and cats. But he soon grew frustrated. Consciousness, he decided, was such a complex topic that it couldn't be fully illuminated in the confined setting of a laboratory. To understand it required an interdisciplinary approach. So, he gave up his animal experiments and immersed himself instead in a broad array of subjects: linguistics, theology, anthropology, neurology, archaeology and literature.

It was from this wide-ranging self-education that he arrived at a revelation. Our consciousness, he realized, must have had an evolutionary history. At some point in the past, our ancestors must have lived immersed in the moment, just like animals, and between then and now our self-awareness developed. Given this, there must have been intermediary steps in its evolution. But what might such a step have looked like? The answer he came up with was that, before our consciousness developed into full-blown self-awareness, it went through a stage in which it took the form of 'voices in the head'. Our ancestors experienced auditory hallucinations that gave them directions.

The way he imagined it was that for most of our evolutionary history, when early humans had lived in hunter-gatherer groups, they got by on pure instinctual behaviours, inhabiting the here and now, focusing on whatever task was at hand. These groups were small enough that, one assumes, they were usually within earshot of each other, and during times of danger everyone could instantly respond to the verbal commands of the leader. As such, there was no need for them to have any kind of introspective self-awareness to regulate their behaviour.

The crucial moment of change, Jaynes believed, occurred approximately 12,000 years ago, when our ancestors gave up the hunter-gatherer lifestyle and settled down to adopt agriculture. This led to the creation of larger communities, like towns and eventually cities, which triggered a problem of social control. People were no longer always within earshot of the leader and certain tasks required them to act on their own. The leader's shouted commands could no longer organize the behaviour of the group.

Our ancestors solved this problem, according to Jaynes, by imagining what the leader might tell them to do. They internalized his voice, and then heard his imagined commands as auditory hallucinations whenever an unaccustomed decision had to be made, or when they needed to be reminded to stay focused on a task. Jaynes offered the example of a man trying to set up a fishing weir on his own. Every now and then, the voice in his head would have urged him to keep working, rather than wandering off as he might otherwise have been inclined to do.

Eventually, the identity of the voices separated from that of the group leader. People came to believe instead that they were hearing the voice of a dead ancestor, or a god. Jaynes theorized that, for thousands of years (until around 1000 BC), this was the way our ancestors experienced the world, relying on these inner voices for guidance.

He detailed this theory in a book with the imposing title *The*

Origin of Consciousness in the Breakdown of the Bicameral Mind.
Despite sounding like the kind of work that only academics could
possibly wade through, its sensational claim caught the public's
interest, and it soon climbed onto bestseller lists.

But why voices in the head? Where did Jaynes get this idea? One
source of inspiration for him was so-called split-brain studies. Our
brains consist of a right and a left hemisphere, connected by a
thick cord of tissue called the corpus callosum. During the 1960s,
surgeons began performing a radical procedure in which they sev-
ered the corpus callosum as a way to treat extreme cases of epilepsy.
This operation left the patients with what were essentially two
unconnected brains in their head.

The procedure did ease the epileptic seizures, and the patients
appeared outwardly normal after the operation, but, as researchers
studied the patients, they realized that at times they behaved in
very strange ways, as if they had two separate brains that were at
odds with each other. For instance, while getting dressed, one
patient tried to button up a shirt with her right hand as the left
hand simultaneously tried to unbutton it. Another would strenu-
ously deny being able to see an object he was holding in his left
hand.

These results made researchers realize the extent to which the
two hemispheres of our brain not only act independently of each
other, but also focus on different aspects of the world. The left
brain is detail-oriented, whereas the right takes in the bigger picture. The left could be described as more rational or logical,
whereas the right is more artistic or spiritual.

It was these split-brain studies that led Jaynes to theorize that
the primitive consciousness of our ancestors might have been
similarly split in two, just as their brains were. For them, the right
hemisphere would have served as the executive decision-maker,
ruminating upon long-term planning and strategy, while the left
hemisphere would have been the doer, taking care of activities in
the here and now. To return to the example of constructing a

fishing weir, the left hemisphere would have handled the minute-by-minute details of the task, while the right hemisphere would have acted as the overall manager, making sure the job got done.

So, most of the time, the left hemisphere would have been in charge, but occasionally a novel situation would have arisen and the left hemisphere would have hesitated, not sure of what to do. At which point, the right hemisphere would have issued a command to it, and the left brain would have experienced this as an auditory or visual hallucination.

Jaynes called this hypothetical form of split-brain mental organization the 'bicameral mind', borrowing the term from politics, where it describes a legislative system consisting of two chambers, such as the House of Commons and the House of Lords.

Of course, you and I also have brains with two hemispheres, but most of us don't hear voices in our heads. Jaynes's hypothesis was that our ancestors simply hadn't yet learned to coordinate the two hemispheres in order to produce what we experience: a singular, unicameral consciousness.

Split-brain studies weren't Jaynes's only source of inspiration. He claimed that we didn't need to try to guess how our distant ancestors experienced reality because we actually had a surviving first-hand account of life during the era of the bicameral mind. By examining this source carefully, we could discern the character of their mental world.

The account he was referring to is one you might have read. It was the *Iliad*, the great epic poem of ancient Greece. It describes events during the Trojan War, which historians believe occurred around 3,200 years ago. We don't know exactly when the *Iliad* was written down, though it was hundreds of years after the war itself. The narrative, however, seems to have been based upon oral tales that were passed down from the time of the war itself. For this reason, Jaynes argued, it could give us a glimpse into the mental world of those people who had lived over 3,000 years ago. It was an artwork, he said, produced by a bicameral consciousness.

When Jaynes read the poem with this goal of trying to reconstruct the mentality of its ancient authors, it quickly became apparent to him that they possessed a very different view of reality than we do. For a start, the characters in the poem displayed absolutely no self-awareness. They never paused for moments of introspection or contemplation. Jaynes noted that the Greek in which the *Iliad* was written didn't even have words for concepts such as consciousness, mind or soul.

Even more intriguingly, whenever the human characters needed to make a decision, they didn't. Unfailingly, a god would appear and tell them what to do. Jaynes detailed examples of this. When Agamemnon takes Briseis, the mistress of Achilles, a god grabs Achilles by his hair and warns him not to raise his sword against Agamemnon. Gods lead the armies when they go into battle and then urge on the soldiers at each turning point. It's the gods who cause men to start quarrelling. In fact, it's the gods who are responsible for the war in the first place. As Jaynes put it, 'the gods take the place of consciousness.'

Classical scholars before Jaynes had noted this peculiar role played by the gods in the poem, but they had always interpreted it as a kind of literary device. Jaynes challenged this assumption. What if, he asked, the gods in the *Iliad* weren't intended to be fictional, but were an actual description of reality as ancient people experienced it? What if they really did hear and see gods giving those commands? After all, he noted, the characters in the poem treat the gods matter-of-factly, as if they were genuinely present.

Of course, he conceded, the Greek and Trojan warriors weren't really seeing gods; they were experiencing auditory and visual hallucinations, but they wouldn't have been able to make this distinction. For them, with their bicameral consciousness, the gods would have seemed very real. Jaynes stated his case bluntly: 'The Trojan War was directed by hallucinations, and the soldiers who were so directed were not at all like us. They were noble automatons who knew not what they did.'

*

In Jaynes's chronology, the Trojan War took place towards the end of the era of the bicameral mind. The transition to modern consciousness began soon after.

The cause of this transition, as Jaynes argued, was that the bicameral mind was too rigid to respond well to truly novel situations. So, as communities grew ever larger and began to bump up against neighbouring groups, signs of tension emerged. A more flexible method of regulating behaviour was needed. In essence, the world of these ancient humans grew more complex, and they required a more sophisticated form of brain organization to deal with it. The end result was the development of modern self-awareness. Instead of waiting to hear the hallucinated commands of a god, people developed an internal 'I' that could make decisions.

Jaynes emphasized that this development didn't involve a physical change to the brain. Modern consciousness was a learned adaptation, a socially acquired skill – one that was then taught to succeeding generations. Even today, Jaynes noted, consciousness is something we learn as children, achieving full self-awareness only around the age of seven. One of the unusual aspects of the brain, in fact, is that it's an extremely malleable organ that develops in response to, and guided by, social contact. Without it, the brain atrophies, such as in the rare cases of feral children raised without human interaction, who permanently lose the ability to learn language and, apparently, to achieve rational consciousness.

But, according to Jaynes, the bicameral mind didn't disappear completely. He argued that it left its mark in many different ways, such as through religion, much of the history of which consists of people hearing voices in their head that they believe to be gods. The prophet Moses, for example, took directions from a voice that supposedly emanated from a burning bush.

Jaynes also proposed that remnants of the bicameral mind endure into the present in the form of schizophrenia. Those who suffer from this condition continue to hear voices in their head, which they often interpret to be those of gods, demons or angels.

Whereas in the ancient world such voices conferred advantages, in the era of unicameral consciousness the voices have become an active hindrance, a medical condition in need of treatment.

Jaynes's fellow scholars weren't quite sure what to make of his theory. Some of them dismissed it as ridiculous. Others were more ambivalent. Writing in 2006, the biologist Richard Dawkins commented, 'It is one of those books that is either complete rubbish or a work of consummate genius, nothing in between! Probably the former, but I'm hedging my bets.'

Scholars offered a variety of criticisms of Jaynes's thesis. Classicists noted that, although most of the *Iliad* is consistent with his theory, not all of it is. When Hector decides to accept the challenge of Achilles, he does appear to engage in introspective contemplation. Psychologists, on the other hand, argued that his suggestion that schizophrenics might experience a kind of bicameral consciousness seemed unlikely, because their hallucinations are far more complex and varied than the type he described. Christopher Wills, a biologist at University of California, San Diego, pointed out that modern-day hunter-gatherers who live isolated lifestyles appear to have the same form of consciousness as everyone else, but, if Jaynes was right, this shouldn't be the case, given their lack of contact with the outside world.

For the most part, however, Jaynes's hypothesis has simply been ignored by scholars. Due to its highly interdisciplinary nature, it seems to fall through the cracks, rarely getting cited. It occupies its own oddball academic niche.

Jaynes does, however, have a few ardent supporters who are convinced he may be onto something. As evidence of this, they point to some findings that have recently emerged from the study of prehistoric cave art.

Our Stone Age ancestors who lived in Europe 10,000 to 30,000 years ago left behind fantastic artwork in caves, mostly depicting the animals in their environment. The sophistication of this art has led most researchers to conclude that they must have possessed minds much like ours. But, in 1999, the cognitive scientist Nicholas Humphrey published an article in the *Journal of Consciousness Studies* in which he argued that this wasn't necessarily the case. Just as Jaynes had found hints in the *Iliad* of a mentality very alien to our own, Humphrey discerned similar peculiar qualities in cave art.

In particular, Humphrey drew attention to the curious similarities between the prehistoric art in Chauvet Cave, in France, and that of an autistic girl named Nadia, born in 1967, in Nottingham. Nadia was an artistic savant. At the age of three, she began producing remarkably accomplished drawings, even though she had received no instruction at all and, indeed, almost entirely lacked language skills. Her drawings, however, displayed a highly distinctive style. Her subject of choice was animals, particularly horses, and she frequently mixed together parts of animals to produce chimera creatures. She had a marked preference for side-on views, and she emphasized faces and feet while largely ignoring much of the rest of the body, often drawing figures haphazardly, one on top of another. This style seemed to stem from her autistic tendency to focus obsessively on individual parts, while ignoring any larger context.

Humphrey noted that the cave art displayed a very similar style, such as the focus on animals, chimera creatures and the seemingly haphazard overlap of drawings. So, while cave art has frequently been offered as evidence of the emergence of the 'modern' mind, he argued that the exact opposite might be true. It might instead be revealing how strangely pre-modern the minds of those ancient painters were.

Humphrey didn't cite Jaynes, but his conclusion strongly echoed the bicameral hypothesis. It offered another hint that our ancestors may have perceived reality in ways profoundly unlike the way we do. Their brains may have been biologically much the same as ours, but they could have been organized along very different principles.

If this was the case, it raises the intriguing thought that our mentality can change. The brain might be able to reorganize itself to navigate reality in new ways. This could have happened before, when it switched from a bicameral to a modern consciousness, and if it's happened in the past, it might occur again in the future. If so, one has to wonder, what strange new form would it assume?

Weird became plausible: beer before bread

Around 12,000 years ago, our ancestors gave up their nomadic lifestyle as hunter-gatherers and settled down to become farmers. It was a pivotal moment in our history as a species, because agriculture led directly to civilization and so to the modern world. But why did our ancestors make this change? This question has long puzzled scientists. Research suggests that hunter-gatherers enjoyed a pretty good life. They had abundant leisure time, and their diet was healthy and varied. Being a farmer, on the other hand, was back-breaking work and the diet was monotonous, which led to problems of poor nutrition and disease. In other words, agriculture didn't seem to improve the standard of living for most people. So, what inspired them to embrace it?

The obvious answer is food. Agriculture would have provided Neolithic humans with a steady supply of grain, which can easily be stored for long periods of time and later used to bake bread. The security of this was surely preferable to the constant threat of not being able to find any food. A less obvious answer, however, is beer. After all, grains can be used to make either bread or beer. Perhaps our ancestors were lured by the appeal of intoxication, and they started planting crops with the intent of becoming brewers rather than bakers.

This is known as the beer-before-bread hypothesis. When it was first proposed in the 1950s, scholars treated it as a bit of a joke.

The idea seemed too silly to take seriously. In recent decades, however, it's been steadily gaining respectability, to the point that the idea can no longer be dismissed as a humorous 'what if?' Many researchers now regard it as quite possible that beer created civilization.

The hypothesis debuted in 1953 following the discovery by archaeologist Robert Braidwood of evidence of Neolithic grain cultivation at a site in Iraq. Braidwood argued in a *Scientific American* article that changing climatic conditions had made it easier for people to grow grains in the region, and so, he concluded, the production of bread must have been the driving force behind their decision to give up hunting and live in sedentary farming villages. But University of Wisconsin botanist Jonathan Sauer promptly challenged this assumption. What if, Sauer asked, making beer was the purpose of cultivating the grain?

To Braidwood's credit, he didn't dismiss Sauer's suggestion out of hand. In fact, he confessed to being quite intrigued by the idea, and he conceded that the evidence didn't clearly support one hypothesis over the other. He had found the remains of cereal grains, as well as the tools for planting and reaping them, but there was no hint of what people were doing with the grains. So, Braidwood decided to put the question to a panel of experts from anthropology and archaeology. Does it seem more likely, he asked them, that our Neolithic ancestors adopted agriculture to make bread or beer? Their responses appeared in an issue of *American Anthropologist*.

Sauer was given the chance to make his case first. He argued that growing and gathering grain would have been an extremely time-consuming process for our ancestors, with the tools they had available. Would they have gone to all that trouble, he asked, for bread? Surely beer would have been more worth the effort. He also pointed out that archaeologists had consistently found wheat and barley grains in combination at Neolithic settlements. This seemed to him like the ingredients for beer rather than bread.

Most of the experts, however, were sceptical. The Danish archaeologist Hans Helbæk joked that it seemed a bit like proposing that early humans domesticated cows in order to make alcoholic beverages from their milk. The botanist Paul Mangelsdorf was even more doubtful of the idea. If people were spending all that time raising crops to make beer, he wondered, what were they eating? Or were they just drinking all the time? He asked contemptuously, 'Are we to believe that the foundations of Western Civilization were laid by an ill-fed people living in a perpetual state of partial intoxication?'

The general conclusion was that the production of some kind of gruel or porridge had probably preceded both beer and bread, since this can be made simply by pouring water on grains. Eventually our ancestors would have figured out that, by cooking the gruel, they could transform it into bread. The brewing of beer, the experts decided, must have followed later.

And that seemed to settle the matter. After the symposium, the beer-before-bread hypothesis disappeared from sight. Scholars proceeded to assume that the transition to agriculture had been a sober affair.

This remained the consensus for thirty years, until the 1980s, when two researchers at the University of Pennsylvania, Solomon Katz and Mary Voigt, revived the case for beer. They noted that, during the intervening decades, archaeological evidence had weakened the case for the bread-first hypothesis. Studies were finding that, for several thousand years after the initial cultivation of grains, Neolithic people had continued to consume a wide variety of plants. This suggested that the decision to take up agriculture had been driven by a cultural desire, rather than by a biological need for food. These early societies were using the grain for something they wanted rather than required. That sounded more like a case for beer than bread.

Simultaneously, evidence from the field of human nutrition had strengthened the beer-first hypothesis. Research had revealed that fermentation is an excellent way to unlock the nutritional content of grains. It converts them from a relatively low- to a high-nutrition food by adding lysine, improving the B-vitamin content and allowing essential minerals to be absorbed more easily. Neolithic beer would also have been calorie dense and full of soluble fibre – completely unlike the thin, filtered lagers one finds in supermarkets today. Plus, the alcohol content would have killed bacteria, making it safer to consume than gruel. It would even have had medicinal value, because it naturally contains the antibiotic tetracycline, produced during the fermentation process. Overall, beer drinkers might actually have enjoyed a significant evolutionary advantage over those who chose to abstain. And this didn't even factor in the pleasurable intoxicating effect provided by the beer.

Given these new findings, Katz and Voigt argued, it was quite plausible to imagine that the discovery of fermentation had been the trigger that prompted early people to start purposefully planting grains.

In the twenty-first century, more support for the beer-before-bread hypothesis has come from several lines of evidence. Anthropologist Patrick McGovern of the University of Pennsylvania has been using biomolecular analysis to examine the residue lining ancient pottery shards. This allows him to determine what was once stored in the pots. Sure enough, more often than not, it was fermented beverages. He's been able to determine that a pot found at the Godin Tepe archaeological site, near the Iran–Iraq border, dating back 5,500 years, contained a barley-based beer.

And, coming at the issue from a completely different perspective, cultural anthropologist Brian Hayden of Simon Fraser University has argued that we shouldn't underestimate or trivialize how much our ancestors liked to party. Having gatherings, then as now, served very basic social needs. It bonded communities together, which would have been advantageous from an evolutionary perspective. And most might agree that parties are usually better with beer than without it.

Hayden notes that potlatches, or elaborate ceremonial feasts, are known to play an important role in the cultures of many tribal people, so he imagines that Neolithic people would have often held feasts to demonstrate their wealth and power to their neighbours. He refers to this as 'competitive feasting'. In this context, beer might have been seen as a high-value food item that contributed greatly to the festivities. It would have been something that, for social reasons, our ancestors would have been very motivated to produce. Bread, on the other hand, simply wouldn't have offered similar cultural rewards.

All these arguments for the beer-before-bread hypothesis have earned it intellectual respectability, although it hasn't quite gained the status of academic orthodoxy. The evidence for it remains circumstantial. But then, so does that for bread first. We'll probably never know for sure what the truth is, but it's quite possible that our ancestors were brewers before they were bakers. Katz and Voigt summed up the case for it this way: imagine that you were a Neolithic person and you could have had either gruel, bread or beer with your meal. What do you think *you* would choose?

What if Homer was a woman?

Throughout history, society has been organized rigidly along lines of gender. The prevailing belief, virtually unchallenged until the twentieth century, was that men produced all things to do with high culture (the arts, politics and sciences), whereas women produced nature (i.e. babies). Men reigned supreme in the public sphere, while women governed the domestic sphere. This was regarded as the natural order of things.

This deep-rooted gender division was reflected in the Western literary canon, which is the list of authors held up as the greatest exemplars of European culture. Until well into the twentieth century, the list consisted entirely of men, including authors such as Milton, Shakespeare, Chaucer, Dante, Virgil, Sophocles and Homer. The genius of these writers, whose works countless generations of students had to learn, was supposed to offer reassuring proof that men really were the superior producers of culture.

For over 2,500 years, Homer was the one constant presence on the list. His two epic poems, the *Iliad* and the *Odyssey*, were widely believed to be not only the greatest, but also the earliest literary works ever written in a European language. They were the foundation upon which Western civilization was built. His cultural influence was immeasurable.

For this reason, Homer's masculinity was taken for granted. No one questioned it. It would have seemed absurd to do so. It went unchallenged until the very end of the nineteenth century, when the British novelist Samuel Butler first put forward the heretical idea that perhaps, for all those centuries, everyone had been wrong. Perhaps Homer actually was a woman.

Surely such an idea can be dismissed easily. There must be some direct evidence of Homer's gender. After all, doesn't he identify himself as a man somewhere in his poems? Don't contemporaneous writings refer to him as such?

As it turns out, no. There's no direct evidence that Homer was a man. Homer never mentioned his/her gender. In fact, the two poems attributed to Homer offer no details at all about their authorship. Their creator remains completely anonymous throughout both, never stepping forward to take credit. The name Homer doesn't even appear anywhere in them. Nor is it mentioned in any contemporaneous source.

The origin of the poems themselves is similarly shrouded in mystery. These are the few facts, such as they are:

Around 1200 BC, a war took place around Troy, which was a city located on the north-west corner of what is now Turkey. Or perhaps it was a series of battles. Archaeologists aren't sure, though they're fairly confident that some kind of conflict took place there, and that it was fought between the residents of Troy and forces from the mainland of Greece. Four hundred years then passed, and in around 800 BC the Greek alphabet was invented. And sometime relatively soon after this, two epic poems about the conflict were set down in writing. They were among the earliest works written in the new alphabet. The first poem was the *Iliad*, set during the battle of Troy itself. The second poem was the *Odyssey*, which followed the adventures of the hero Odysseus as he tried to return home after the war, but, on his journey, was waylaid by monsters and sorceresses.

Once the poems had been written down, however, even more time passed before the Greeks started to wonder about who had written them. The identity of the author was lost in the mists of history. It was now after 500 BC, and, in the absence of any firm information, a legend sprang up attributing the poems to an ancient bard named Homer. The Greeks, though, were well aware that they didn't actually know anything about who Homer was, and even more legends sprang up to fill the biographical void about him. The most popular of these was that he had been a blind poet who had lived on the coast of Asia Minor.

This meagre suite of details about the Homeric poems has left many questions unresolved. Scholars are sure that the poems must have existed to some degree in oral form before they were written down, recited by bards during public festivals. But were they composed in their entirety as oral poetry, perhaps soon after the Trojan War, and then passed down for many generations before finally being transcribed? Or did the person who wrote them down essentially invent them, perhaps weaving together fragments of older oral poems? Scholars simply don't know.

Also, precisely how soon after the invention of the Greek alphabet (circa 800 BC) were the poems written down? The classicist Barry Powell has argued that the alphabet itself may have been invented in order to write them down. Others, however, put their composition as late as 600 BC.

Scholars aren't even sure Homer was just one person. Many believe the works we attribute to Homer were composed by lots of different authors and then stitched together at some later date. Another theory suggests that the name Homer referred to a guild of poets.

Homer's biography, in other words, is a blank slate. It's only tradition that ascribes masculinity to the poet.

However, pointing out the lack of biographical information about Homer merely establishes uncertainty. Is there any reason to truly suspect Homer was a woman?

Samuel Butler thought so. He detailed his case for this in a 270-page book titled *The Authoress of the Odyssey*, published in 1897. His argument focused entirely on the *Odyssey*, which he felt had a distinctly feminine sensibility to it. So much so that he concluded it could only have been written by a woman.

Butler noted that the *Odyssey* was full of female characters: Penelope, Minerva, Eurycleia, Helen, Calypso, Circe, Arete and Nausicaa. In fact, it contained more women than any other ancient epic, and these women were not only central to the plot, but they were also fully fleshed out and sympathetic. The male characters seemed to Butler to be rather wooden and unintelligent.

There were other details that, to his discerning eye, betrayed a woman's hand, such as the expert way in which domestic life was portrayed in the epic. He also listed various errors in the *Odyssey* that, he believed, only a woman would have made. He called them a 'woman's natural mistakes'. These included: 'believing a ship to have a rudder at both ends' and 'thinking that dry and well-seasoned timber can be cut from a growing tree'. Surely no man would have made such obvious blunders! So, of course, the author of the *Odyssey* must have been a woman.

Butler even managed to narrow down who this female author was. He identified her as a young unmarried girl who lived in the town of Trapani, in north-west Sicily, around 1050 BC – a conclusion he arrived at by identifying landmarks around Sicily that resembled locations described in the *Odyssey*. Just about the only thing he didn't provide was a street address.

Butler's argument may seem distinctly odd to us. After all, a 'woman's natural mistakes'? Really? Plus, his reasoning only applied to the *Odyssey*. He thought the *Iliad*, with its violent battle scenes, was probably the work of a man. He believed there were two separate Homers, one of whom was a woman, and that, by some historical accident, both poems had been credited to one man.

Nevertheless, his book landed like a cultural H-bomb, sending shockwaves of controversy far and wide. It didn't really matter that his argument wasn't very good. The mere fact that he had dared to make it was scandalous. Scholars were outraged. This was the

Victorian era, when the British educational system was still centred around learning the classics. Homer was like the training manual for the male upper-class elite. To question his gender was to challenge the entire patriarchal order upon which British society was built.

But Butler found ardent supporters among modernist artists, who, in the early twentieth century, were busy trying to challenge all kinds of taken-for-granted notions, such as what art is supposed to look like. Why couldn't a toilet or a blank canvas count as art? some of them were asking. So, the Homer-was-a-woman theory was right up their alley and became a cause célèbre among them. Its most famous legacy was that it inspired James Joyce to write his masterpiece *Ulysses*. Joyce reportedly kept a copy of Butler's book beside him on his desk as he wrote.

To give Butler his due, the *Odyssey* is curiously female-centred. If an ancient woman had sat down to write an epic, it's easy to imagine she would have produced something like the *Odyssey*. But, if that were the entirety of the evidence for the female-Homer theory, it wouldn't be very interesting, except as a historical curiosity. But there's more to it, thanks to the efforts of historian Andrew Dalby, who, in his 2007 book *Rediscovering Homer*, updated Butler's argument, offering a more sophisticated case for why Homer might have been a woman.

Dalby lay the groundwork for his argument by establishing some historical context. He noted that ancient Greece had plenty of talented female poets who could have produced the works of Homer. Poetry was an art form that Greek women not only participated in, but excelled at. Sappho, who lived in the sixth century BC on the island of Lesbos, was widely regarded as being one of the greatest Greek poets.

Dalby also dismissed the widely held belief that epic poetry was performed only by men, with women restricted to singing lamentations or love poetry. Studies of surviving traditions of oral poetry have shown, he said, that both men and women composed and performed epic poetry. The difference was that only men performed it in public settings, such as banquets, while women performed it in private, family settings.

Finally, Dalby rejected the idea that the epic poems could have been composed in oral form before the invention of the Greek alphabet and written down later. This, he maintained, simply wasn't the way oral poetry worked. Poems were never transmitted verbatim from one generation to the next. The idea of such perfect transmission is a concept that belongs to the print mindset. In oral culture, only a very loose framework of ideas and repeating phrases was ever passed down. Each performance was essentially a new version of the story. In which case, whoever first wrote down the Homeric poems must essentially have created them anew.

With this context in mind, Dalby then asked the question, why would a male bard have written down these epic poems? It seems like a very strange thing for a man to have done. After all, today we take it for granted that writing is a way to gain cultural status, but in Ancient Greece, circa 800 BC, hardly anyone could read. Writing was a brand-new technology. Writing down epic poems would have been a dubious venture, at best. Who would have been the intended audience?

Here was the crux of Dalby's argument. He believed that a successful male poet (and the author of the *Iliad* and the *Odyssey* must have been *the* best poet in Greece) might have resisted participating in such a novel undertaking because there was no obvious status to be gained from it. Honour and glory for male bards came from public performance. A written text was an unknown quantity. Plus, participating in such a project – presumably reciting the poetry in a private setting while a scribe wrote it down – would have taken him away from his career, possibly for months.

A female poet, on the other hand, would have had no qualms about performing in a private setting. If she was a talented poet, barred from performing at public gatherings, she might even have sensed the possibility of reaching an entirely new kind of audience through writing.

Dalby also noted that the anonymity of the poems might support their female authorship. Male poets traditionally identified themselves in their works and bragged about their accomplishments; it was an important part of the process of crafting a public persona. The lack of any identification in the Homeric poems hints, therefore, that their author was someone in the habit of maintaining a more private persona: someone such as a woman.

In trying to reconstruct the process by which the poems came into existence, Dalby theorized that the *Iliad* must have been written first by this unnamed female poet. He surmised that she would have had a male patron who bankrolled the entire venture. The goatskin parchment on which the poems were written would, after all, have been very expensive. At his request, she initially created a traditional epic, weaving together old stories and legends.

Twenty years later, Dalby imagined, the same patron must have asked her to write a second epic, but she was now older and had been able to think more deeply about the craft of writing and its possibilities. So, this time, she produced a more experimental, complex work that was also more outwardly feminine. In this way, we gained the *Odyssey*.

Dalby knew that his theory was highly speculative, but he insisted that it was entirely plausible.

Actually, classical scholars haven't rejected Dalby's hypothesis out of hand. Reviewers praised his argument as interesting and imaginative, but they still weren't quite ready to accept the idea that Homer might have been a woman. They pointed out that, even if it's possible that women sang epic poems in private, there's still a very long tradition that identifies epic poetry as a male genre. And, while the author of the Homeric epics may not identify him/herself, there are several bards who are alluded to in the epics, and they're all men. A female Homer, one would think, might have included a female bard in the poems, just for the sake of gender solidarity.

There was also the expense of the Homeric project. Writing

down the epics would have required a significant investment. It wasn't as if goatskin parchment was being mass produced at the time. Would a wealthy male patron really have entrusted such a costly venture to a woman?

Finally, there's that vast weight of tradition identifying Homer as a man. Surely this has to count for something.

Classicists readily acknowledge that none of these are slam-dunk arguments against the female-Homer theory. One gets the idea that, unlike their Victorian-era counterparts, they'd actually really like to believe that Homer was a woman. They just can't bring themselves to.

Though, one has to acknowledge that Dalby does have a point. The Homeric poems came into existence at the exact historical moment of transition from an oral to a written culture. Writing would prove to be the ultimate disruptive technology, eventually completely eclipsing oral culture. Would an accomplished male bard, proud of his success, really have helped usher in the change that was going to spell the end of his profession? That would be a bit like imagining the print and music industries had immediately embraced the Internet. In reality, they long resisted it. It tends to be those on the margins, the outsiders who aren't benefiting from the status quo, who first spy the full potential of disruptive technologies, and, in ancient Greece, who would have been more of an outsider than a woman who possessed all the talent of her male peers, yet could share none of their public glory?

What if Jesus was a mushroom?

Mind-altering substances have probably been used in religious worship for as long as it has occurred. This claim isn't particularly controversial. Ancient Hindu texts refer to the use of an intoxicating drink called soma, and Native American tribes have been using the psychoactive peyote plant in rituals for thousands of years. Archaeologists have found that temples throughout the Mediterranean, such as the Temple of Apollo at Delphi, were often located at sites where trance-inducing hydrocarbon gases generated by bituminous limestone rose out of geological fissures.

Such practices may seem rather exotic to modern-day Christians for whom Sunday worship typically doesn't get any headier than having a sip of sacramental wine. But, according to John Marco Allegro's sacred-mushroom theory, detailed in his 1970 book *The Sacred Mushroom and the Cross*, Christianity wasn't always so tame. In its original form, he claimed, it was very much associated with the use of mind-altering substances. In fact, he claimed that it started out as an ancient sex and drug cult. Even more controversially, he made the case that Jesus was never imagined by the first Christians to be an actual person. Jesus, he said, was a hallucinogenic mushroom.

If Allegro had been a long-haired, wild-eyed radical standing on a street corner, his theory could have been easily dismissed. The problem was, he wasn't anything like that. His academic credentials were impeccable. He was a specialist in philology, the study of ancient languages. He had studied at Oxford University under Sir Godfrey Driver, the foremost figure in the field, and he held a lectureship at Manchester University. The real feather in his cap, however, was his position as a member of the international research team tasked with investigating and translating the Dead Sea scrolls.

The Dead Sea scrolls have been described as the greatest archaeological discovery of the twentieth century. Found in the late 1940s, they are a set of writings that had been hidden in a cave near Qumran in Israel by an ancient Jewish sect, about a century before the time of Jesus. These scrolls contained what were by far the earliest known copies of the Old Testament. To be part of the team researching them was, for all involved, a sign of great academic status. Allegro had been appointed to the team at the urging of Sir Godfrey, who believed him to be a rising star in the field of philology.

By 1970, Allegro had become a well-known public intellectual through his work on the scrolls. He had written several bestselling books, and he appeared frequently on radio and TV. This meant that, when he started declaring that Jesus was a mushroom, his fellow academics couldn't just ignore him. He clearly was in a position to speak authoritatively about early Christianity.

The cornerstone of Allegro's theory was linguistic analysis. He claimed that having a knowledge of ancient languages other than Greek and Hebrew opened up all kinds of novel insights into the New Testament. Particularly important was Aramaic, the language spoken in the ancient world by many Semitic people from the Near East, and Sumerian, the language of ancient Mesopotamia. Scholars had only recently deciphered the latter.

Armed with his expanded linguistic knowledge, Allegro had reread the New Testament and he believed that, in doing so, he had discovered two distinct levels of meaning within it. There was the surface meaning told by the Greek text, which detailed the story of a kindly preacher named Jesus who spread a gospel of love. Beneath this, however, he claimed to have found a level of meaning that only became apparent if one knew Aramaic and Sumerian. This consisted of a complex web of wordplays, puns and allusions that repeatedly referenced mushrooms.

There was, for example, the story of Boanerges. In the Gospel of Mark, two brothers named James and John are introduced, whom, we are told, Jesus had nicknamed Boanerges. Mark tells the reader that Boanerges means 'sons of thunder'. Except, Allegro pointed out, it doesn't – not in any known dialect of Hebrew, Aramaic or Greek. 'Sons of thunder' was, however, an Aramaic term for mushrooms, because they appeared in the ground after thunderstorms.

Another cryptic fungi reference involved the name of Peter, the leading apostle of Jesus. *Petros* means 'rock' in Greek, which serves as the occasion for a famous instance of wordplay in the New Testament when Jesus says, 'Now I say to you that you are Peter, and upon this rock I will build my church.' Allegro noted there was an even deeper pun because *Pitra* was the Aramaic word for a mushroom.

Then there was the scene when Jesus, as he hung on the cross, cried out before his death, *'Eloi, Eloi, lama sabachthani.'* This is usually translated as, 'My God, my God, why hast thou forsaken me?' Allegro informed his readers, *'lama sabachthani* is a clever approximation to the important Sumerian name of the sacred mushroom "LI-MASh-BA(LA)G-ANTA".' *Eloi, Eloi,* he further revealed, recalled the invocatory chant of Bacchic revellers, *'eleleu eleleu',* which they repeated as they pulled the sacred fungus from the ground.

These were just a few of the hidden mushroom references that Allegro found in the New Testament. There were many more. But what should we make of them? They seemed far too numerous to chalk up to chance or coincidence. Allegro concluded that they were, instead, evidence that the authors of the holy book must have been members of some kind of mushroom cult.

Once clued into its existence, Allegro began to suspect that this cult had played a pivotal role in the history of Western religion – a role which had previously been undetected by scholars and was only stumbled upon by him because of his linguistic expertise. He became obsessed by the idea of uncovering this history. His daughter, Judith Anne Brown, later wrote that her main memory of him from this time, when she was a teenager and he was working on *The Sacred Mushroom and the Cross*, was that he was forever in his office, barricaded behind trays of index cards and reams of notes, compiling evidence to support his theory.

The story, as he eventually pieced it together, was that, long ago in the Stone Age (he was vague about the exact date), a mushroom-worshiping fertility cult must have arisen among the people living in Mesopotamia. He imagined these people believing that rain was the sperm of the great sky god that fell on the womb of the Earth, giving birth to her children, the plants. Mushrooms, to them, would have been the most special plant of all, because the fungi possessed remarkable powers. They grew from the soil as if by magic, without a seed – a virgin birth – and their shape resembled a thrusting phallus, making them a symbol of the fecundity of the sky god manifested on Earth.

One mushroom in particular, Allegro theorized, would have captured the attention of these primitive people. This was the psychedelic species *Amanita muscaria*, more commonly known as fly agaric. It has a very distinctive appearance, featuring a bright red cap speckled with white dots. The really attention-grabbing aspect of this mushroom, however, wasn't what it looked like, but rather what happened when it was ingested, because it then granted wondrous visions. The early Mesopotamians might have believed these were glimpses of the divine knowledge of the sky god himself.

Allegro concluded that a cult must have formed around the

worship of this mushroom, with a priesthood dedicated to preserving the various rituals, incantations and preparations necessary for using it. As this was powerful knowledge that couldn't be shared with the common people, the priests made sure it was kept secret, hidden behind an elaborate veil of cryptic codes and secret names, and never written down.

So far, so good. Allegro's theory of an ancient Mesopotamian mushroom cult may have been seen as highly conjectural by other scholars, but few would have taken offense at it. His next move, however, proved far more controversial, because he skipped ahead in his alternative history of religion to the first century AD.

During the intervening centuries, he speculated, the mushroom cult had clashed repeatedly with the secular authorities of monarchs and their bureaucrats. The kind of visionary powers its priests invoked were far too volatile for the liking of political leaders. It had been forced to go underground, but pockets of it remained – small groups who continued to pass down the sacred knowledge of the mushroom from one generation to the next. One of these groups, he concluded, was to be found in the Roman province of Judea. It was a radical Jewish sect whose members called themselves Christians.

The term Christian, Allegro explained, derived from the Greek word *Christos*, meaning 'rubbed on' or 'anointed'. That much is standard linguistics. Jesus Christ means 'Jesus the anointed one'. The traditional explanation is that the term refers back to the ancient Jewish practice of anointing kings with oil. Allegro, however, hypothesized that the Christians might have acquired the name because they rubbed a mushroom-infused hallucinogenic oil into their skin as part of their rituals. And that was just the start of his reinterpretation of Christian symbolism. His other claims were even more sensational.

For example, Allegro claimed that the name 'Jesus' didn't refer to a person at all. It was actually an ancient term for the fly agaric mushroom, deriving from the Hebrew name Joshua, which, in turn, came from a Sumerian phrase meaning 'semen which saves', referring to the mushroom as the phallic representation of the sky god.

Christians also frequently spoke of the death and resurrection of Jesus. Again, according to Allegro, this was part of mushroom lore, being a coded allusion to the life cycle of the mushroom: how it sprouted quickly, as if from a virgin birth, spread out its canopy, then died, but returned again in a few days. Similarly, he said, the cross that the Christians used as the symbol of their faith was fungal in origin. The Aramaic verb for 'crucify' meant 'to stretch out'. So, the 'crucifixion' of Jesus referred to the stretching out of the mushroom to its fullest extent. The cross itself was simply a highly stylized representation of the mushroom with its canopy stretched out. 'The little cross', Allegro noted, was an Aramaic term for a mushroom.

All this might have remained hidden knowledge, according to Allegro, if the Romans hadn't sent troops to Judea in 66 AD to put down the Jewish unrest there, sparking a brutal war that dragged on for seven years and culminated with the famous Siege of Masada. He speculated that the Romans must have cracked down particularly hard on the Christians who had been stirring up trouble by proclaiming that the end of times – and, by extension, the end of the Roman Empire – was near. This was a belief they had arrived at, he said, thanks to their mushroom-induced visions.

Forced to flee Jerusalem, Allegro speculated that the Christian priests had decided on the desperate gamble of writing down their sacred lore, lest it be lost entirely. They took elaborate measures to conceal their secrets, however, embedding them within a cover story about a friendly Jewish rabbi who preached universal love. This text was the New Testament, which brought Allegro's alternative history of religion full circle, as it provided him with an explanation for the cryptic mushroom references which had been the puzzle that launched his scholarly journey of discovery in the first place.

There was one more act in Allegro's alternative history, however. It was an ironic twist of fate, which occurred, he said, when knowledge of how to interpret the Christian texts properly was forgotten during the following centuries. This left the cover story as all that people understood, and it was on this misleading facade that modern Christianity arose.

Although, he said, one could still see echoes of the original mushroom cult in various Christian practices, such as the ritual of Holy Communion. Didn't it seem odd, he asked, for Christians to participate in an apparently cannibalistic ritual involving the consumption of the body and blood of their god? Of course it was! The original ceremony, he assured readers, was far less grisly. It was merely the communal act of sharing and ingesting their god, the sacred mushroom.

The Sacred Mushroom and the Cross appeared in bookshops in May 1970. The British media heavily promoted it, unable to resist the spectacle of such heretical claims being argued by a well-known academic. On account of this publicity, it sold briskly, until readers discovered that, despite the book's scandalous theme, it was no page turner. In fact, it was almost unreadable to anyone but a philologist. Allegro, believing he was writing a work of great intellectual significance, had earnestly tried to make it scholarly enough to satisfy his fellow academics, to the point of including 146 pages of endnotes analysing arcane points of linguistics.

He probably shouldn't have bothered, because scholars savaged the book. A typical review came from the theologian Henry Chadwick, Dean of Christ Church, Oxford, who wrote that it read like 'a Semitic philologist's erotic nightmare'.

Critics zeroed in on Allegro's philological argument, in which he claimed etymological links between Greek, Aramaic and Sumerian words. These presumed links seemed absurd to them. After all, just because words and phrases may sound vaguely alike – such as *'lama sabachthani'* and *'LI-MASh-BA(LA)G-ANTA'* – that doesn't mean they're related, as Allegro assumed. Peter Levi, reviewing the book for the *Sunday Times*, noted that it also didn't make sense to believe that a Jewish sect in the first century AD would have been proficient in Sumerian.

Even sympathetic reviewers, such as the poet Robert Graves, found it difficult to accept Allegro's argument, questioning whether *Amanita muscaria* was even growing in Judea at that time in order to have been available to the cult members.

The hardest blow for Allegro, however, landed when *The Times* published a letter signed by fifteen of the leading linguistic professors in the United Kingdom, who declared that the book wasn't based on any philological evidence they considered worthy of scholarly significance. Among the signatories was Sir Godfrey Driver, Allegro's former professor at Oxford.

As Allegro's daughter later noted in her biography of him, the book ruined his career, and, to a large degree, his life. His marriage fell apart. He lost his job and never held an academic position again. He continued writing books, though for an increasingly small audience, and, in 1988, at the age of sixty-five, he dropped dead of an aortic aneurysm.

A book that's gained so much infamy may seem impossible to defend – but is it? Allegro's philological argument is probably beyond salvage. Even his most loyal defenders in the twenty-first century, such as his daughter, concede that his linguistic speculations are unconvincing. But what about his more general claim that Christianity originated from a mushroom cult? As outlandish as the idea might seem, an argument could potentially be made for that – although, to do so, one has to be willing to accept a rather large premise: that Jesus never existed as a historical person.

As far as mainstream scholars are concerned, the debate ends right there. The firm consensus is that Jesus, putting aside the question of whether he had any supernatural attributes, was a real person. But, for over a century, a handful of scholars have been insisting there are compelling reasons to doubt the historicity of him. This is called the Christ-myth theory, and its advocates are known as mythicists.

They note that there's no direct evidence of Jesus's existence. All the information about his life comes from sources written decades after he supposedly lived. That alone isn't particularly significant; the same is true of many people throughout history whose existence isn't doubted. But the case of Jesus is more suspicious, they maintain, because there's a remarkable similarity between his life story and the stories of various beings worshipped by pagan cults.

During the Greco-Roman era, cults had sprung up throughout the Mediterranean whose followers worshipped divinities such as Osiris, Romulus, Adonis and Mithra. There was a common pattern among these beings. They were all said to be of human form, though born of a divine parent, and they had all undergone some form of suffering through which they had obtained a victory over death, enabling them to offer personal salvation to their followers. Moreover, although these beings were divine, they were all said to have lived here on Earth, and stories were told about them set in human history.

By the first century AD, this pattern had become well established. So, the mythicists argue, isn't it possible that the story of Jesus was simply an outgrowth of this broader pagan trend? Might not Jesus have originated as a mythic being similar to the other ones worshipped around the Mediterranean, and over time his story began to be presented as fact, embellished with real-world details?

But, if Jesus was a mythic being, mainstream scholars respond, why does the story of his life end with his crucifixion? That was a highly stigmatized way to die in the Roman Empire. Surely a cult would have chosen a more respectable death for their deity. The fact that they didn't, most scholars believe, suggests that this detail was based on a real-life event. It indicates that there really was a preacher named Jesus, and this was how he died.

Despite this objection, the Christ-myth theory nevertheless endures with a robust, if fringe, following. And, for those willing to proceed down this path, the question then becomes, what type of mythic being was Jesus originally?

Most mythicists believe he had been a deity associated with the sky or sun. But this is where Allegro's theory returns, because couldn't Jesus just as well have been a mushroom deity? If the argument is over what type of mythic being he was, why is a mushroom god any less plausible than a sky god? It would, after all, explain why early Christians so frequently described having mystical visions. And there were indisputably ancient cults organized around the worship of drug gods. The most famous of these was the cult of Dionysus, the Greek god of wine. Like Jesus, Dionysus was believed to have undergone death and resurrection.

So, that's one possible argument for taking Allegro's theory seriously. Though, the fact remains that it does seem rather outrageous to transform the historical Jesus into a mushroom, and this actually highlights a central feature of Christianity: what a crucial role the apparent historicity of Jesus has played in its success. A god that seemed less flesh-and-blood and more mythical wouldn't have had the same persuasive power. So, even if, for the sake of argument, one is willing to consider that Jesus may have originated as a mushroom, the more relevant observation would be how completely this knowledge was erased – because the reason Christianity became one of the dominant religions in the world was precisely because so many people were convinced that Jesus was entirely real.

Weird became true: ancient Troy

German-born businessman Heinrich Schliemann liked to tell the story of how he first heard about Troy. It was, he said, in 1829, when he was only seven, and his father gave him a copy of Ludwig Jerrer's *Illustrated History of the World* for Christmas. In it, he found a picture of the ancient city of Troy in flames. Intrigued, he showed it to his father, who took him on his knee and told him about the legendary beauty of Helen and of how her love for the Trojan prince Paris led her to flee her husband, King Menelaus; how the Greeks had launched a thousand ships to reclaim her from Troy, and how the heroes Achilles and Hector fought on the dusty plains outside the city; and finally of how the Greeks used trickery to gain access to the walled citadel by hiding inside a wooden horse. Then they burned Troy to the ground.

Schliemann claimed that hearing this tale stirred something inside of him, making him decide, even at that young age, that one day he would find Troy. But, at the time, most historians didn't believe that Troy existed. Throughout ancient history, its existence had been taken for granted, but with the rise of more sceptical attitudes towards historical analysis in the eighteenth century, doubts had crept into the minds of scholars. After all, there were no obvious remains of anything like a city where it should have been located, on the north-west corner of Turkey. This scepticism acquired orthodox status when the classical historian George

Grote published his multi-volume *History of Greece* in the mid-nineteenth century. Grote argued that there was no more reality to the tale of the Trojan War, with its larger-than-life heroes, than there was to other ancient legends, such as the story of Jason and the Argonauts. Poetry, he declared authoritatively, was not history. There was simply no reason to believe the city was real.

Schliemann thought differently, but he was still a young man when Grote's history came out, so he put his dream of finding Troy on a back burner – although he would later claim not a day passed when he didn't think of it – and he devoted himself instead to business. He became a commodities trader, and he proved to be brilliant at it. He had a natural talent for languages, which facilitated his dealings in the world of international business. By the early 1860s, when he was still only in his forties, he had become so wealthy that he didn't have to work another day in his life.

He set off to travel the world, and, in 1868, he found himself in the eastern Mediterranean. It was time, he decided, to finally pursue his dream of finding Troy once and for all. He had already taught himself ancient Greek some years ago, so, with a copy of Homer's *Iliad* in his hand, he visited sites in north-west Turkey, comparing landmarks with Homer's description of the landscape around Troy.

Homer had written that, from Troy, you could see the snow-capped peak of Mount Ida, that there were two rivers there, and that it was close enough to the coast so that the Greek warriors could easily walk from their camp on the beach to the city itself. By following these clues, Schliemann eventually decided that a site named Hisarlik, near the modern seaport of Çanakkale, had to be Troy.

Schliemann was so confident in his identification that, before he even began digging, he published his conclusion as a book called *Ithaca, the Peloponnesus and Troy*. Academics weren't convinced. The French historian Ernest Renan declared him a fool. Others sneered that he was a mere businessman. What, they asked, could he possibly know?

Schliemann remained undeterred. Their scepticism simply made him more determined to prove his claim. In 1870, he hired a team of diggers and launched a full-scale excavation at Hissarlik. They dug a trench, forty-five feet deep, straight through the hill at the site, revealing that a powerful city definitely had once been located there. In fact, the excavation showed that the city had been built up and destroyed nine different times. These layers were stacked horizontally on top of each other.

Schliemann's greatest achievement came in May 1873, when he discovered what he declared to be the 'Treasure of Priam' (in the Homeric epics, Priam was the King of Troy). It was a stunning collection of bronze, silver and gold artefacts that included jewellery, battle axes, swords, shields and vases. It certainly seemed like a treasure worthy of ancient Troy. The find made headlines around the world. Together with the other evidence from the dig, it seemed obvious, at least to the popular press, that Schliemann had proven the experts wrong. He had discovered ancient Troy.

Scholars still refused to believe his claim for years. Continuing excavations over the course of the twentieth century, however, led most of them to agree that Hisarlik probably is Troy. There's no smoking gun, such as a throne with Priam's name on it, but the site is in the right location, it was a wealthy city, and there's evidence of armed conflicts there at the end of the Late Bronze Age, which is when historians believe the Trojan War would have occurred. So, the thinking goes, why wouldn't it be Troy?

This would seem to vindicate Schliemann on the question of Troy's existence, but, despite this, his reputation hasn't actually fared that well. In fact, it's pretty much in tatters, because it turns out that the story just told, of his lifelong dream to find Troy and how he defied the cynics to do it, is itself a mixture of myth and reality. As historians began to look more closely at his life, examining the diaries he kept compulsively, they concluded that he was an inveterate liar, given to self-aggrandizement. Yes, he did excavate at Hisarlik even though classical scholars insisted Troy was a myth, but it has since become clear that he stole the credit for the discovery of the city from a mild-mannered British expatriate named Frank Calvert.

The reality seems to be that Schliemann, despite what he often claimed, never had a childhood dream of finding Troy. That story was pure invention, part of the myth he constructed around himself following the discovery. He had only developed an interest in archaeology during the early 1860s, after retiring from business, when he was experiencing a midlife crisis, searching around for something to do that would earn him cultural prestige to match his wealth. He initially flirted with the idea of becoming a writer, but then he fastened upon archaeology after attending some lectures on the subject at the Sorbonne. It wasn't until he happened to meet Calvert in Turkey in 1868 that he fixated upon Troy as his life's great work.

It was Calvert, not Schliemann, who had nursed an obsession for years of finding Troy. Long before Schliemann showed up in the eastern Mediterranean, Calvert had already identified Hisarlik as the probable site of Troy. He had even purchased half the site to acquire the digging rights, and he then tried to interest the British Museum in financing a dig there, but they turned him down. He had started excavating on his own, but he had made little progress due to a lack of free time and funds.

When Schliemann showed up in Turkey in 1868, Calvert thought it was his lucky day. Here was a wealthy businessman interested in archaeology! Calvert shared everything he knew about Hisarlik, and Schliemann eagerly listened.

To Schliemann's credit, he instantly realized the opportunity Calvert had handed him, and he seized it. But, having no further need for Calvert, he proceeded to bulldoze him to the side. Calvert lacked the temperament or resources to fight back effectively. As a result, by 1875, Schliemann was basking in the public glory of being the discoverer of Troy. He published another book, *Troy and its Remains*, triumphantly recounting his work at the excavation. In it, he didn't even mention Calvert at all.

The discovery of Troy does, therefore, offer an example of academic experts being proven wrong, but not quite in the inspirational way you would hope. Instead of being a lesson about pursuing your dreams even if cynics say they're impossible, the moral of the story is more like, be careful who you share your dreams with, because a rich businessman might steal them and take all the credit when they come true.

What if Jesus was Julius Caesar?

Around 33 AD, Roman forces in Judea executed a messianic Jewish preacher named Jesus. For them, the execution was a minor event. Certainly, none of them would have predicted that his death might have any long-term impact on their empire. But, of course, that's exactly what happened. His followers, the Christians, fanned out across the Mediterranean, preaching their leader's message to whomever would listen, and their numbers grew at a remarkable rate, eventually displacing the pagan religions of the empire altogether. Today, approximately one third of the population of the entire world is estimated to be Christian – over two billion people.

How did the early Christians manage to pull off this feat? How did an obscure sect founded by a carpenter from the backwoods of Galilee manage to end up wielding such vast global influence and power? According to the Church, it was the bravery of Christian martyrs in the face of persecution that inspired mass conversions among the Roman people to the faith. Secular scholars point to a variety of other factors, such as the ceaseless evangelism and missionary work of the early Christians, as well as the strong cohesive social networks they built.

In the 1990s, an Italian linguist named Francesco Carotta came forward with a far stranger answer. According to him, mainstream historians were starting from the wrong premise. They assumed that Christianity really had begun as an obscure Jewish sect, whereas the reality, he argued, was very different. He believed that Christianity actually originated at the highest level of the Roman Empire, and so it had enormous imperial resources behind it from the very beginning. And how could this be? It was because Jesus wasn't the man everyone thought he was. In fact, he was the product of the greatest case of mixed-up identity in history. Jesus, insisted Carotta, was really Julius Caesar.

The revelation of Jesus's true identity first occurred to Carotta in the 1980s, as he was looking at pictures of statues of Caesar. Although Caesar was a great military leader, who, in the first century BC, almost became Rome's first emperor, Roman artists often portrayed him with a soulful, spiritual expression – one that seemed very Christlike to Carotta. That was when the idea popped into his head. What if the face of Caesar was the original face of Christ?

Carotta became so obsessed by this idea that he eventually sold the software company he had founded and devoted himself full-time to researching the Caesar–Christ connection, drawing upon his previous academic training in linguistics and ancient history. The result was the publication, in 1999, of a massive 500-page book detailing his discovery. The original German title posed his thesis as a question, *War Jesus Caesar?* (*Was Jesus Caesar?*), but the title of the 2005 English translation was more definitive: *Jesus Was Caesar*.

Carotta's argument rested principally on a fact that's well known to classical historians, but which isn't well known by the public: Caesar was worshiped as a god. A group of Roman senators killed Caesar in 44 BC by stabbing him to death, claiming they did so to stop him from transforming the Roman Republic into an empire, with himself at its head. His assassination failed to stop the imperial transformation of the Republic, however. It merely delayed it for seventeen years, until Caesar's adopted son Octavian (subsequently given the title Augustus) succeeded in his place, becoming the first emperor. But, meanwhile, Caesar's followers had promptly declared their fallen leader to be a divinity, equal in rank to Jupiter, the supreme god of the Roman pantheon. They built temples and monuments around the Mediterranean to honour 'Divus Julius', the 'Divine God Julius'. Caesar became the first Roman leader ever elevated to the status of a god, although not the last. Subsequent Roman emperors, such as Augustus, were also considered to be gods.

This cult of Caesar worship, Carotta contended, was the original form of Christianity. All the symbols and stories that we associate with Christ emerged there first. Divine Caesar was swapped out for Divine Jesus.

Carotta built his case, first and foremost, on a series of linguistic and narrative similarities between the lives of Christ and Caesar. Whatever was in the life of Christ, he insisted, could be found earlier in the life of Caesar, as if someone had merely tweaked Caesar's biography, changing names and details slightly, to transform it into the story of Christ.

The most obvious similarity was the identical initials, J. C., shared by the two men. Then there was the fact that Caesar rose to power as a general in Gaul (or Gallia, to use the Latin spelling), whereas Jesus began his ministry in Galilee. Gallia and Galilee. A coincidence? From Gallia, Caesar travelled south to a holy city (Rome), where he was killed by Roman enemies who feared that he desired to become a king, and afterwards he ascended to heaven as a god. Likewise, Jesus travelled south from Galilee to a holy city (Jerusalem), where he was killed by Romans who claimed he desired to become a king (the King of the Jews). Afterwards, he too ascended to heaven as a god.

The senator that delivered the fatal dagger wound, killing Caesar, was Gaius Cassius Longinus. And, according to a legend that dates back to the early centuries of Christianity, as Christ hung on the cross, a Roman centurion named Longinus stabbed him in the side with a spear.

Both Caesar and Christ had a traitor. Brutus betrayed Caesar, and Judas betrayed Christ. The names aren't similar, but Carotta noted that Brutus's full name was Decimus Junius Brutus. Junius, his family name, is written in Greek as 'Junas', which is similar to Judas.

Then there were the tales of miracles attributed to Jesus, which, Carotta argued, closely resembled the tales that circulated of Caesar's various miraculous feats during military campaigns. In the writings of the Roman historian Appian, for example, we find a story about Caesar ordering his troops to cross the sea by Brundisium at night. To help his men, he slipped incognito into the boat with them, but the crossing grew rough, and his men grew afraid. This prompted Caesar to reveal himself and say, 'Do not fear, you sail Caesar in your boat, and Caesar's luck sails with us!' They made the crossing safely.

This seems oddly similar to the tale of what is perhaps Jesus's most famous miracle, when he walked on water, which was said to have occurred as his disciples were trying to cross the Sea of Galilee at night. Caught in a storm, they feared for their lives, when suddenly they saw Jesus walking on the water, through the waves. He approached them and said, in words like those spoken by Caesar, 'Be of good cheer: it is I; be not afraid.' And his disciples made the crossing safely.

Carotta dug up many more similarities between Christ and Caesar. His ultimate goal, in fact, was to trace everything in the gospels back to the life of Caesar. All these possible parallels, however, seem to overlook the very basic difference between the two men, which was that Christ was a poor carpenter who preached a message of peace, whereas Caesar was a ruthless, powerful general. How could the lives of two such diametrically opposed figures possibly have got mixed up?

The process began, according to Carotta, with the rise of the Caesar cult, because it contained within it many of the elements that would later be associated with Christianity.

For a start, there was Caesar himself. While he was indeed a general, Carotta pointed out that he also had a reputation among the Roman people as a champion of the poor, celebrated for his compassion and charity. These were the same qualities later attributed to Jesus.

Thousands of people attended Caesar's funeral to mourn the loss of the great leader, and Carotta argued that a curious theatrical device featured at the service laid the basis for the Christian symbol of the cross. The device was a cross-like apparatus, attached to which was a wax figure of Caesar showing him as he lay in death, arms spread out. This was then lifted upright so the entire crowd could witness Caesar's wounds. The sight of it was said to have so inflamed the crowd with grief that they rose up en masse and spilled out into the streets of Rome to hunt for Caesar's murderers.

After Caesar's funeral, his cult spread throughout the empire, which at the time stretched from northern France all the way down to Egypt and encompassed the entire Mediterranean region. Both his son Octavian and Mark Antony, Caesar's former right-hand man, promoted the worship. It took root, in particular, among the veterans who remained intensely loyal to the memory of their former leader. He had won their allegiance by granting them plots of land throughout the empire in return for their service. As the retired veterans settled on this land, they served as the evangelists who disseminated the faith. According to Carotta, they took with them a holy text – a biography of Caesar written by one of his followers, Asinius Pollio, soon after the general's death. Carotta imagined the veterans using this text as a kind of Gospel of Caesar, learning from it stories of Caesar's virtue, compassion and miraculous life.

Carotta noted that, if these army veterans were the first evangelists of the faith, it might explain why Christians began referring to non-believers as 'pagans'. The word derives from the Latin *pagus* meaning 'village'. The veterans' camps were usually located on the outskirts of villages, so it would have been natural for the retired soldiers to differentiate between themselves, worshipers of Caesar, and the local villagers (pagans) who weren't.

And the veterans as evangelists might solve another mystery: why early Christian writers showed a marked preference for the codex format, as opposed to continuous scrolls. A codex is the scholarly term for a book made from stacked and bound sheets of paper. Which is to say, it's the style of almost all books today (not counting e-books!). The first codices began appearing around the first century AD, and Christians were immediate and enthusiastic adopters of the format, whereas non-Christian writers persisted in using scrolls for centuries.

Scholars aren't sure why this was the case. Why did Christians care whether they wrote on codices or scrolls? To Carotta, the reason was obvious. It was because the inventor of the codex was widely believed to be none other than Julius Caesar. The story goes that, during military campaigns, he started folding his scrolls, concertina style, finding them easier to read that way, and this subsequently inspired the creation of the codex. Therefore, it would have made perfect sense for his followers, the veterans, to prefer this format. It was a way to imitate and honour him.

In Carotta's revisionist history, many of the elements of Christianity, such as the cross, holy text and a community of worshipers, were now in place. The religion had also spread throughout the empire. Caesar, however, remained the central figure of worship. How did he get switched out for Jesus? Carotta believed this involved deliberate deception. It was a plot concocted by the Roman emperor Vespasian, around 75 AD, with the help of a Jewish historian in his court, Flavius Josephus.

Before becoming emperor, Vespasian had risen to power as the general who put down the Jewish uprising in Judea – a long, bloody war that took place between 66 and 73 AD. After the conflict, Carotta speculated, Vespasian sought for a way to integrate the Jews into the empire, to ensure they wouldn't revolt again. From the Roman point of view, the problem with the Jews was their religious zeal, so Vespasian's idea was to weaken their religiosity by converting them into emperor worshipers. He recognized, however, that in order for this to happen, the emperor cult had to be translated into a form that would resonate with them. It had to be Judaized. Vespasian tasked Josephus with the job of implementing this plan, and he duly set to work.

His strategy was to create a Jewish version of the cult of Caesar. He used Pollio's biography of Caesar as a base, and he basically took the story of Caesar's final days and transplanted it into Judea, replacing Caesar with an imaginary Jewish preacher named Jesus. The text that resulted from these transformations, claimed Carotta, was the Gospel of Mark. Historians believe this gospel was written around the time of the Jewish War, so the timing is about right for Josephus to have authored it. And here we encounter yet another of Carotta's curious links between Christianity and Caesar, because scholars don't actually know who the Mark was who wrote the Gospel of Mark. It's always been a mystery. But it wasn't to Carotta. He believed it was a reference to Mark Antony.

With this gospel in hand, Josephus then set out to promote the new religion, using the veterans' camps as his base of operations. His refashioned version of Caesar's life turned out to be a smash hit, and the rest is history. Josephus himself, Carotta claimed, would later have his own identity transfigured, turned into none other than the Apostle Paul, the Roman Jew who played a pivotal role as a leader of the early Church.

Is that it, then, for Christianity? Will its followers have to acknowledge that they're really Caesarians? Will Christmas have to be changed to Caesarmas?

Carotta's theory *has* gained a small but passionate group of supporters who evangelize on its behalf. With their help, Carotta's book and articles have been translated from German into a variety of languages: Dutch, English, Italian and French. But, of course, Christianity is in no imminent danger. Negative reactions to Carotta's theory far outnumber the positive ones. Critics have denounced it as pseudoscience, eccentric, a bad joke, the work of a charlatan, utter lunacy and (best of all) 'monkey cabbage'. No academic journal has ever even bothered to review his book.

The most frequent criticism is that the similarities Carotta finds between Caesar and Christ are trivial and most likely accidental. And, really, his detractors ask, how could the cult of Caesar have transformed into Christianity without eliciting comments from anyone in the classical world? It seems mind-boggling that no one would have mentioned anything. And, by reimagining Christianity as a Roman cult that was Judaized, rather than a Jewish sect that was Romanized, Carotta simply brushes aside the huge body of scholarship that has mapped the deep connections between the ministry of Jesus and the Jewish culture of the first century AD.

So perhaps Carotta's theory can be dismissed as ridiculous and absurd. And yet, in its defence, if you were to strip his argument down to its essence (ignoring the part about Caesar actually being Jesus), what he's suggesting is that there's a deep link between emperor worship and Christianity. He imagines that, without the former, the latter may never have arisen. Phrased in this way, his theory isn't that crazy. In fact, it's close to being in line with a great deal of recent scholarship.

For a long time, scholars assumed that emperor worship wasn't a true religion – it was mere politics, little more than an elaborate sham and a kind of imperial contrivance worked up by Rome to force its provinces to show deference.

These assumptions began to be challenged in the 1980s due to the work of Oxford historian Simon Price, who found evidence that emperor worship was very much a sincere form of belief. He demonstrated that the emperor cults were typically grass-roots movements that sprang up spontaneously throughout the empire, rather than being top-down creations of Rome. Emperor worship, Price argued, provided people with a way to come to terms with the empire itself. From their point of view, the emperor genuinely had as much power over their lives as a god, so it seemed natural to them to worship him as one would a divine being.

But, if emperor worship was an authentic form of belief, this complicated the story of its relationship to Christianity. Instead of Christianity filling a vacuum created by the collapse of paganism, scholars began to wonder if it may have arisen on top of a base supplied by emperor worship. The emerging consensus is that, in fact, this seems to be the case – that the imperial cults heavily influenced early Christians, who adopted a great deal of vocabulary and symbolism from them.

New Testament scholar Bart Ehrman, for example, has argued that it isn't mere coincidence that Christians claimed their crucified leader to be a god just around the same time that the Romans were referring to their dead emperors as gods. The example supplied by emperor worship of how a man could turn into a god would have been hard to ignore. Similarly, Fordham University professor Michael Peppard has pointed out that the Emperor Augustus, the adopted son of Julius Caesar, was officially referred to as the 'Son of God', the identical title later applied to Jesus.

It makes sense that the Christians would have incorporated elements from emperor worship into their own doctrine, even if only to contrast their own faith with it. After all, this was the symbolic language that potential converts around the Mediterranean were familiar with.

Viewed from this perspective, all those parallels that Carotta found between Caesar and Jesus don't seem quite as implausible. Mainstream scholars definitely wouldn't agree that Jesus was actually Caesar, but could the Christians have borrowed elements from the cult of Caesar, adapting many of its stories to embellish the story of their messiah? The idea is controversial, certainly, but possible. In which case, Jesus wouldn't *be* Caesar, but his biography might contain echoes of the fallen general's life within it.

What if the Early Middle Ages never happened?

Try this simple test. Name something that happened in Europe between 614 and 911 AD.

If you're a history buff, it's probably an easy question. After all, there are records of many different events from that period. There was the rise of the Carolingian dynasty in the eighth century, which culminated with the reign of Charlemagne the Great, or there were the Viking raids that spread terror throughout much of Europe in the ninth and tenth centuries.

If you managed to name some things, congratulations. But if you couldn't, don't worry. Most historians wouldn't ever say so, but, according to the German scholar Heribert Illig, 'nothing' would be the most accurate answer. Illig is the author of the phantom-time hypothesis, according to which the 297 years between 614 and 911 AD never took place – they were conjured out of thin air and inserted into European history. He believes that the events which supposedly occurred during this time – the rise of Charlemagne and all the rest – were total fabrications.

As Illig tells it, it was in the late 1980s when he first stumbled upon the idea that approximately 300 years had been added to the calendar. Although he had earned a doctorate in German language and literature studies, he was supporting himself as a systems ana-lyst at a bank. It evidently wasn't his dream job, but he was making ends meet. Then, one day, he was thinking about the Gregorian calendar reform of the sixteenth century when he realized that there was a mystery lurking within it – one that led him to the idea of phantom time.

In 1582, Pope Gregory had ordered a reform of the calendar. The problem was that the Julian calendar, established by Julius Caesar in 45 BC, was growing out of sync with the seasons. The Julian year was 365¼ days long, but this was 674 seconds shorter than the actual solar year. Over the span of centuries, these few seconds had started adding up. The Pope realized that, if something wasn't done soon, Easter would soon be celebrated in winter rather than spring. His scholars, at his behest, determined that ten days needed to be skipped to get the calendar back in line with the solar year, and this is what the Pope decreed should be done.

Here, as Illig saw it, was the mystery. Why was it only necessary to skip ten days? Every 128 years, the Julian calendar drifted an additional day apart from the solar year. This meant that, by 1282, a ten-day error would have accumulated, which was the amount of days the Church skipped. But, of course, they reformed the calendar in 1582, not 1282. There was an additional 300 years unaccounted for. They should have had to skip thirteen days, not ten. Why the discrepancy?

Illig decided there was only one possible explanation. At some point between 45 BC and 1582 AD, approximately 300 years must have been added to the calendar.

Illig realized that what he was contemplating would be regarded as outrageous by conventional scholars, but he nevertheless continued to follow this train of thought. If three centuries had been added to the calendar, he asked himself, which centuries were most likely to be the fake ones? They would probably suffer from poor documentation, he reasoned, and exhibit an overall lack of source material. After all, a forger could never match the richness of genuine history. Historians would probably regard those centuries as being particularly obscure and little understood. Was there such a period in European history? It immediately occurred to him that there was indeed. It was the centuries that made up the bulk of the era known as the Dark Ages.

The Dark Ages is a vague term. Its temporal boundaries aren't strictly defined. It's generally applied to the period from the end of the Western Roman Empire in 476 AD to around 1000 AD. Modern historians don't like the term at all, thinking it unnecessarily judgemental. They prefer to call these centuries the Early Middle Ages. But the term Dark Ages has stuck in popular usage because the period does seem dark – we don't know a lot about it due to the meagre amount of source material that survives. It was also a fairly grim period of European history, with the darkest period lasting for about 300 years, from 600 to 900 AD.

It's as if, during these centuries, European civilization fell off a cliff. A handful of barbarian kingdoms had risen up in place of the Roman Empire and, while their rulers initially tried to keep Roman laws and customs in place, maintaining the facade that they were preserving the tradition of imperial order, as time passed, things crumbled. Infrastructure stopped being built. The great engineering projects of the Romans fell into ruin. Literacy rates plummeted, not that they were very high to begin with, and knowledge was lost as traditions of scholarship eroded.

There's been much speculation about what caused this decline. Theories have attributed it to the disruption of Mediterranean trade by Islamic pirates, or even to enormous volcanic eruptions in Central America that could have triggered climate change, impacting agriculture in Europe. But most historians now chalk it up to the effects of disease. A series of epidemics swept through Europe, beginning in the third century, and, as more time passed, these epidemics kept hitting. Population levels fell dramatically. There simply weren't enough people to maintain civilization at its former levels, so much of the population of Europe reverted to a more primitive style of existence.

But, Illig asked, what if the real reason for the decline was that it was all an illusion? What if the Dark Ages were dark because they were fake?

In this way, Illig recast the obscurity of those centuries as a sign of chronological distortion, rather than of disease and low population. The truth, he said, was that, after the fall of the Western Roman Empire, Europe had experienced only a slight bump in the road before resuming a steady ascent to the economic and cultural achievements of the High Middle Ages. He singled out the years 614 AD and 911 AD as the boundaries of the falsehood. The first date corresponded to the Eastern Roman Empire's loss of Jerusalem to the Persians, while the latter date marked the treaty between the Viking duke Rollo and Charles the Simple, King of West Francia. Both these events, he decided, had really happened, but everything in between, including the rise of the Carolingian dynasty, he dismissed as an invention.

How could such an outrageous chronological manipulation possibly have taken place? Such an elaborate scheme must have had an architect. Who was it, and why would they have done it?

Illig's suspicions focused on a youthful ruler, Emperor Otto III, who was part of the Ottonian dynasty that established a vast kingdom throughout Germany and northern Italy in the tenth century. According to conventional history, Otto III was born in 980 AD and became king when he was only three, following the death of his father, Otto II. But as he grew into manhood, young Otto developed lofty ambitions. He not only wanted to rule the largest kingdom in Europe, he also wanted to lead what he called a *renovatio imperii Romanorum* or a 'renewal of the empire of the Romans'. His plans, however, were cut short when he died of fever at the age of twenty-one.

Illig offered a slightly different version of Otto's life. He preserved the broad outline of the ruler's biography, but placed him in an earlier century. By Illig's reckoning, Otto was born in 683 AD instead of 980 AD. He still saw him as a young man of brash ambition, though. In fact, Illig speculated that his desire may have extended far beyond the political sphere, into the spiritual as well. Illig imagined that Otto wanted to play a leading role, not just in the affairs of men, but in the great divine drama of the universe itself. He yearned to be Christ's representative on Earth, who would ring in the final millennium of creation before the arrival of the Day of Judgement.

Christian doctrine at the time held that the world would exist for 7,000 years. Each 1,000-year period was thought to correspond to a single day in God's time, analogous to the seven days of creation. By the calculations of seventh-century scholars, they were currently living in the sixth millennium. There were different estimates of when the seventh, final millennium would begin, but the year 1000 seemed to many like a good bet. After all, the Book of Revelation had foretold that Satan would stay bound for 1,000 years, which could be interpreted to mean that he would return 1,000 years after Christ's birth, and that this would mark the commencement of the final millennium, triggering an epic struggle between the forces of good and evil leading up to the end of the world.

These weren't fringe ideas. This was the orthodox, official teaching of the Church. There was fervent anticipation throughout the Christian world for the arrival of the final millennium. Without a doubt, Otto III shared these beliefs. But Illig believed that Otto faced a problem in fulfilling his great ambition because he had been born too soon, in 683, which meant that the final millennium was still over 300 years off.

So, Illig speculated, the idea might have formed in Otto's mind that the date of the final millennium didn't lie hundreds of years in the future, but that it was actually close at hand. This is a common psychological phenomenon among members of end-of-the-world cults. They always want to speed up the arrival of the apocalypse. They convince themselves that they're witnessing the end times.

If Otto truly believed this, then the actual date on the calendar would have been a minor inconvenient detail. He could simply change the date to match his fevered, millennial fantasy. After all, he was emperor! Perhaps he convinced himself that he was correcting the date, rather than altering it. Or perhaps he didn't come up with the idea on his own. Pope Sylvester II might have whispered the idea in his ear to flatter the young ruler's vanity. Some form of collusion between the Pope and Otto would have been necessary to pull off the scheme.

Having decided on this grand plan, Otto would have sent forth couriers to deliver the order to clerics throughout his kingdom: add 300 years of history to the calendar! Obediently, they would have bent over their desks to begin working. We don't need to imagine they would have done so unwillingly. After all, they probably shared the emperor's millennial expectations; it was the mindset of the era, all part of the fervent hope to witness the unfolding of Biblical prophecy in the present.

Such a scheme might sound wildly implausible. Surely even an emperor couldn't have engineered so outrageous a deception! But this, Illig countered, is to think like a person of the twenty-first century. In the modern world, it would be impossible to surreptitiously add 300 years to the calendar, but, in the seventh century, it would not only have been possible, it would have been easy.

Most people in the Early Middle Ages had no idea what the date was. That information was irrelevant to them. Only a handful of clerics and scribes knew how to read and write, and only they cared about the calendar. Otto could have changed the date, and it wouldn't have caused the smallest ripple in the lives of most people in his kingdom. Vast culture-wide indifference worked in his favour.

Just as importantly, in the seventh century, almost no one used the *anno Domini*, or AD, dating system, so there would have been no resistance to altering it. At the time, it was still most common to date events by referring to the ruler in power. One would say, for example, that such-and-such an event had occurred in the fifth year of the reign of Otto III.

The idea of using Christ's birth as year zero had been introduced in around 525 by a Scythian monk named Dionysius Exiguus, but it was slow to catch on. It only began to gain in popularity around the time of the Ottonian dynasty, and even then its adoption was so gradual that it wasn't until 1627 that it occurred to anyone to use its counterpart, BC, or 'before Christ'. According to Illig, it truly wouldn't have been difficult for a suitably determined emperor to have manipulated the *anno Domini* calendar. In fact, he argued, the reason we use the AD dating scheme today is precisely because the phantom-time plotters promoted its use.

Illig published his hypothesis in 1991, detailing it in a German-language book titled *Das erfundene Mittelalter* (*The Invented Middle Ages*). Or would it be more accurate to say he published it in 1694?

Historians in Germany were incredulous. Illig's claims seemed to them to be so absurd that they scarcely merited a response, and if his book hadn't started to climb bestseller lists, they probably would have ignored it. But the book did attract public attention, so they felt obliged to issue a rebuttal of some kind. But what could they say? How do you prove that 300 years actually happened?

Illig's hypothesis actually raised questions of a truly existential nature for historians, and this is arguably the most interesting aspect of it. They were questions such as, what allows us to say anything with certainty about the past? What is our knowledge about history ultimately based upon?

These kinds of questions are so basic, they normally never get asked outside of dry academic discussions of historical methodology. Illig, however, was raising them in a very public, sensational way, challenging the validity of historical knowledge itself.

So, historians patiently tried to explain the types of evidence that led them to believe the Early Middle Ages really had happened. They noted the existence of archaeological evidence from that period. This included buildings, some of them quite spectacular, such as the Palatine Chapel built for Charlemagne in Aachen around 800 AD. There was also comparative world history. The chronologies of other regions around the world, such as the Middle East and China, meshed seamlessly with European history. How could this be true if three hundred years of the Western calendar were invented?

They regarded the most compelling evidence of all, however, to be the over 7,000 written sources that survive from the Dark Ages. These are internally consistent from one country to another. The information from English chroniclers matches that from French and German ones. For it to have been faked would have required a vast army of monks and clerics engaged in an international conspiracy of historical forgery. Such an idea seemed, on its face, ridiculous.

Historians conceded that there was no single piece of evidence that, on its own, could prove the reality of the Dark Ages. Instead, it was the sum total of all the evidence, each bit supporting the other, that provided a solid foundation for belief.

Illig, however, didn't buy it. He and his supporters challenged each type of evidence. Why not consider the possibility of a vast conspiracy of forgers? they asked. After all, medieval clerics hardly had a sterling reputation for honesty. The modern ideals of historical accuracy simply hadn't yet been developed, back then. For the clerics, the purpose of keeping records was to support the interests of the Church or king. They happily faked the records, if need be.

As for comparative world history, Illig speculated that other cultures had readily incorporated the phantom centuries into their own chronologies, his thinking being that, if ancient rulers were offered a blank historical canvas, they would find some way to fill it. And the archaeological evidence? He dismissed that as misdated.

Faced with these arguments, most historians concluded that further debate was pointless. They adopted an unofficial ban on further discussion of Illig's hypothesis, calling this policy *Totschweigetaktik*, 'death by silence'.

*

But Illig, in his peculiar way, did have a valid point. He was right that historical knowledge isn't absolute. There's always a lingering uncertainty attached to it. This is why there's a tendency to look down upon it as not being as rigorous and objective as knowledge obtained from the experimental sciences, such as physics and chemistry. It's viewed as inherently more speculative and circumstantial. This may be part of the reason why Nobel Prizes aren't awarded for the historical sciences, not even for geology. It's probably also why weird theories flourish in these disciplines, because the evidence is more open to interpretation.

Given this, it is possible to systematically question the validity of every piece of historical evidence. Radical scepticism is an option. In fact, it could be taken even further than Illig took it. In 1921, the philosopher Bertrand Russell posed what has come to be known as his five-minute hypothesis. He noted that the entire world might have sprung into existence five minutes ago, complete with memories of earlier times. So, forget about the Early Middle Ages not existing. How can we even know that yesterday existed?*

Russell argued that we actually can't know this for sure. One could arrive, like René Descartes, at the conclusion that existence itself is the only thing we can know with certainty: 'I think, therefore I am.' Outside of that, all is potentially illusion.

Almost all scholars reject this kind of thinking as an intellectual dead end, which of course it is. It rejects the very possibility of evidence, because in theory it could all be artificially manufactured. And yet, if you want to go that route, there's nothing to stop you. If you're willing to question everything, then you have to admit as a logical possibility that the past never existed.

* This recalls the idea that the universe might be a computer simulation. See 'Chapter One: Cosmological Conundrums'. In fact, the phantom-time hypothesis could be seen as a variant of the simulation hypothesis, with medieval monks having been the architects of a simulated historical memory.

Epilogue

We've covered a lot of ground. We've explored the earliest moments of the cosmos, investigated the origin of life, witnessed the evolution of the human species and have finally arrived at the rise of civilization. Now we can see the dim outlines of the modern world emerging: the formation of nation states, the establishment of the great religions of the world and the creation of universities that will foster scientific and technological progress.

But this is where we'll bring our journey to an end. The intent of this book was to offer a broad introduction to the genre of weird theories. Having sampled from all the great eras of cosmic history, we've achieved that aim, although our investigation could easily continue, as there are numerous odd hypotheses left to explore. There are many that address present-day concerns. What if, for example, global warming is making us fatter? There's a theory that the rise in global levels of carbon dioxide is both heating the planet and stimulating appetite-related hormones in our bodies.

We could also look beyond the present, to the future. What will become of humanity if we survive our careless management of the planet? Some speculate that our destiny is to miniaturize: our descendants will find ways to access the inner space of the microscopic and even quantum realms, until they finally transport themselves into a 'black-hole-like destination'.

And how will the universe itself end? Or will it ever? One

unnerving possibility is that the entire cosmos could undergo a sudden random change in its energy state, much like an atom undergoing the process of radioactive decay. In which case, absolutely everything in existence might abruptly vanish without warning, at any moment.

In fact, the scope of weird theories is as wide as curiosity itself. It encompasses all disciplines and domains of knowledge. Such speculations reflect our restless urge to make sense of this world around us, to discern the hidden connections lurking beneath the surface of reality, coupled with the suspicion that current explanations may not suffice.

Of course, these weird ideas may be no more than mad flights of fancy. They could be sending us on fool's errands, down blind alleys. Most of them probably are. That's a definite risk. But, on the other hand, isn't it also a risk to ignore them entirely? After all, science has repeatedly shown the world to be far stranger than anyone would have expected.

Unfortunately, no algorithm exists that can reliably pick out the hidden gems from among the larger mass of misguided notions. There's simply no substitute for analysis, debate and the constant evaluation (and re-evaluation) of evidence. And, on occasion, ideas that initially seemed absurd do emerge victorious from this process.

I believe this demonstrates one lesson above all: remain curious! Be willing to consider strange ideas that challenge mainstream opinion. That doesn't mean embracing every wacky notion that comes along. Scepticism is important as well. But it does mean that one should never be afraid to ask questions, even seemingly stupid ones. Those are often the very best kind.

Acknowledgements

It would be weird to imagine that authors write books alone. They don't. They need lots of help, and this book was no exception.

I'm deeply indebted to Charlotte Wright at Pan Macmillan who patiently guided the manuscript to completion. Her many suggestions and critiques were invaluable.

My family and friends, meanwhile, helped to keep me sane and focused during the entire process. My parents gave constant encouragement from afar, as did Kirsten, Ben, Astrid and Pippa. Charlie offered comic relief, and kept me humble about my Scrabble skills. Thanks to Danielle for being such a great daughter-in-law, and introducing me to the world of IPAs. Kingston, my first grandchild, arrived just as the manuscript was almost complete, but has already provided much love and many smiles. Pumpkin made sure I was up every morning on time and sat with me every day in my office. Stuart and Caroline provided helpful advice about British terminology. Anne and Diana hosted relaxing Yuma getaways – in addition to other times just spent hanging out. But most of all, thanks to Beverley for lovingly putting up with my long journey through the land of weird theories.

Bibliography

Introduction

Kuhn, Thomas S., *The Structure of Scientific Revolutions* (3rd edn.). Chicago: University of Chicago Press (1996).

Chapter One: Cosmological Conundrums

WHAT IF THE BIG BANG NEVER HAPPENED?

Bondi, H., *Cosmology*. Cambridge: Cambridge University Press (1961).

Gregory, J., *Fred Hoyle's Universe*. Oxford: Oxford University Press (2005).

Hoyle, F., *Steady-State Cosmology Re-visited*. Cardiff: University College Cardiff Press (1980).

Hoyle, F., Burbidge, G. & Narlikar, J. V., *A Different Approach to Cosmology: From a Static Universe Through the Big Bang Towards Reality*. Cambridge: Cambridge University Press (2000).

Narlikar, J. V. & Burbidge, G., *Facts and Speculations in Cosmology*. Cambridge: Cambridge University Press (2008).

WEIRD BECAME TRUE: RADIO ASTRONOMY

Jarrell, Richard, 'Radio Astronomy, Whatever That May Be: The Marginalization of Early Radio Astronomy' in *The New Astronomy: Opening the Electromagnetic Window and Expanding our View of Planet Earth*. Dordrecht: Springer (2005).

Kellermann, K. I., 'Grote Reber (1911–2002)' in *Publications of the Astronomical Society of the Pacific* 116, pp.703–11 (August 2004).

WHAT IF OUR UNIVERSE IS ACTUALLY A COMPUTER SIMULATION?

Bostrom, Nick, 'Are you living in a computer simulation?' in *Philosophical Quarterly* 53(211), pp.243–55 (2003).

Hanson, Robin, 'How to live in a simulation' in *Journal of Evolution and Technology* 7(1) (2001).

Hossenfelder, Sabine, 'No, we probably don't live in a computer simulation' in *Back Re(Action)* (15 March 2017). Retrieved from http://backreaction. blogspot.com/2017/03/no-we-probably-dont-live-in-computer.html.

WHAT IF THERE'S ONLY ONE ELECTRON IN THE UNIVERSE?

Feynman, Richard P., 'The Development of the Space–Time View of Quantum Electrodynamics' in *Science* 133(3737), pp.699–708 (12 August 1966).

Gardner, Martin, 'Can time go backward?' in *Scientific American* 216(1), pp.98–109 (January 1967).

Halpern, Paul, *The Quantum Labyrinth: How Richard Feynman and John Wheeler Revolutionized Time and Reality*. New York, NY: Basic Books (2017).

Schwichtenberg, Jakob, 'One Electron and the Egg'. Retrieved from http:// jakobschwichtenberg.com/one-electron-and-the-egg/

WHAT IF WE'RE LIVING INSIDE A BLACK HOLE?

Carroll, Sean, 'The Universe is Not a Black Hole' (28 April 2010). Retrieved from http://www.preposterousuniverse.com/blog/2010/04/ 28/the-universe-is-not-a-black-hole/

Luminet, Jean-Pierre, *Black Holes*. New York, NY: Cambridge University Press (1992).

Pickover, Clifford A., *Black Holes: A Traveler's Guide*. New York: Wiley (1996).

WEIRD BECAME TRUE: DARK MATTER

Bertone, Gianfranco & Hooper, Dan, 'A History of Dark Matter' in *Reviews of Modern Physics* 90(4) (October–December 2018).

Hooper, Dan, *Dark Cosmos: in search of our universe's missing mass and energy*. New York: Smithsonian Books/Collins (2006).

Rubin, Vera C., 'Dark Matter in the Universe' in *Proceedings of the American Philosophical Society* 132(3), pp.258–67 (September 1988).

WHAT IF WE LIVE FOREVER?

Byrne, P., *The Many Worlds of Hugh Everett III*. New York: Oxford University Press (2010).

DeWitt, B. S., 'Quantum mechanics and reality' in *Physics Today* 9, pp.30–5 (1970).

DeWitt, B. & Graham, N. (eds.), *The Many-Worlds Interpretation of Quantum Mechanics*. Princeton: Princeton University Press (1973).

Lewis, P. J., 'Uncertainty and probability for branching selves' in *Studies in History and Philosophy of Modern Physics* 38, pp.1–14 (2007).

Tegmark, M., 'The interpretation of quantum mechanics: many worlds or many words?' (1997). Retrieved from https://arxiv.org/abs/quant-ph/9709032.

Chapter Two: A Pale Blue Peculiar Dot

WHAT IF THE EARTH IS AT THE CENTRE OF THE UNIVERSE?

Clifton, T. & Ferreira, P. G., 'Does dark energy really exist?' in *Scientific American* 300(4), pp.58–65 (2009).

Davies, P. C. W., 'Cosmic heresy?' in *Nature* 273, pp.336–7 (1978).

Ellis, G. F. R., 'Is the Universe Expanding?' in *General Relativity and Gravitation* 9(2), pp.87–94 (1978).

Ellis, G. F. R., Maartens, R. & Nel, S. D., 'The expansion of the Universe' in *Monthly Notices of the Royal Astronomical Society* 184, pp.439–65 (1978).

WHAT IF PLANETS CAN EXPLODE?

De Meijer, R. J., Anisichkin, V. F. and van Westrenen, W., 'Forming the Moon from terrestrial silicate-rich material' in *Chemical Geology* 345, pp.40–9 (8 May 2013).

Herndon, Marvin J., *Maverick's Earth and Universe*. Victoria, British Columbia: Trafford (2008).

Ovenden, M. W., 'Bode's Law and the Missing Planet' in *Nature* 239, pp.508–9 (27 October 1972).

Van Flandern, Tom, *Dark Matter, Missing Planets and New Comets*. Berkeley, California: North Atlantic Books (1993).

WEIRD BECAME TRUE: THE HELIOCENTRIC THEORY

Gingerich, Owen, *The Book Nobody Read: Chasing the Revolutions of Nicolaus Copernicus*. New York: Walker & Company (2004).

Westman, Robert S., *The Copernican Question: Prognostication, Skepticism, and Celestial Order*. Berkeley, California: University of California Press (2011).

WHAT IF OUR SOLAR SYSTEM HAS TWO SUNS?

Davis, M., Hut, P. and Muller, R. A., 'Extinction of species by periodic comet showers' in *Nature* 308, pp.715–17 (19 April 1984).

Muller, Richard, *Nemesis: The Death Star*. New York: Weidenfeld & Nicolson (1988).

Schilling, Govert, *The Hunt for Planet X*. New York: Copernicus Books/Springer Science (2009).

WEIRD BECAME TRUE: CONTINENTAL DRIFT

Oreskes, Naomi, *The Rejection of Continental Drift: theory and method in American earth science*. New York: Oxford University Press (1999).

Powell, James L., *Four Revolutions in the Earth Sciences: From heresy to truth*. New York: Columbia University Press (2015).

WHAT IF TEN MILLION COMETS HIT THE EARTH EVERY YEAR?

Dessler, A. J., 'The Small-Comet Hypothesis' in *Reviews of Geophysics* 29(3), pp.355–82 (August 1991).

Frank, L. A., Sigwarth, J. B. and Craven, J. D., 'On the influx of small comets into the Earth's upper atmosphere' in *Geophysical Research Letters* 13(4), pp.303–10 (April 1986).

Frank, Louis A., *The Big Splash*. New York, NY: Birch Lane Press (1990).

WHAT IF THE EARTH IS EXPANDING?

Edwards, M. R., 'Indications from space geodesy, gravimetry and seismology for slow Earth expansion at present—comment on "'The Earth expansion theory and its transition from scientific hypothesis to pseudoscientific belief"'' in *Hist. Geo Space. Sci.* 7, pp.125–33 (2016).

Kragh, H., 'Expanding Earth and declining gravity: a chapter in the recent history of geophysics' in *Hist. Geo Space. Sci.* 6, pp.45–55 (2015).

Nunan, R., 'The theory of an expanding Earth and the acceptability of guiding assumptions' in *Scrutinizing Science: Empirical Studies of Scientific Change*. Dordrecht: Kluwer Academic (1988).

Nunan, R., 'Expanding Earth theories' in *Sciences of the Earth: An Encyclopedia of Events, People, and Phenomena*, Vol. 2. New York: Garland Publishing (1998).

Sudiro, P., 'The Earth expansion theory and its transition from scientific hypothesis to pseudoscientific belief' in *Hist. Geo Space. Sci.* 5, pp.135–48 (2014).

Chapter Three: It's Alive!

WHAT IF EVERYTHING IS CONSCIOUS?

Goff, Philip, 'Panpsychism Is Crazy, but It's Also Most Probably True' in *Aeon* (1 March 2017). Retrieved from https://aeon.co/ideas/panpsychism-is-crazy-but-its-also-most-probably-true.

Shaviro, Steven, 'Consequences of Panpsychism' in *The Nonhuman Turn*. Minneapolis: University of Minnesota Press (2015).

Skrbina, David, *Panpsychism in the West*. Cambridge, MA: The MIT Press (2017).

WHAT IF DISEASES COME FROM SPACE?

Hoyle, Fred & Wickramasinghe, N. C., *Diseases From Space*. London: J. M. Dent & Sons (1979).

Hoyle, Fred & Wickramasinghe, N. C., *Evolution From Space*. New York: Simon and Schuster (1981).

WEIRD BECAME PLAUSIBLE: THE VENT HYPOTHESIS

Corliss, J. B., Baross, J. A. & Hoffman, S. E., 'An hypothesis concerning the relationship between submarine hot springs and the origin of life on Earth' in *Oceanologica Acta* 4 (supplement), pp.59–69 (1981).

Hazen, Robert M., *Genesis: The Scientific Quest for Life's Origin*. Washington, DC: Joseph Henry Press (2005).

Radetsky, Peter, 'How did life start?' in *Discover Magazine* 13(11), pp.74–82 (November 1992).

WHAT IF THE EARTH CONTAINS AN INEXHAUSTIBLE SUPPLY OF OIL AND GAS?

Cole, S. A., 'Which Came First, the Fossil or the Fuel?' *Social Studies of Science* 26(4), pp.733–66 (November 1966).

Glasby, G. P., 'Abiogenic Origin of Hydrocarbons: An Historical Overview' in *Resource Geology* 56(1), pp.85–98 (2006).

Gold, Thomas, *Power from the Earth: Deep Earth Gas – Energy for the Future*. London: Dent (1987).

Gold, Thomas, *The Deep Hot Biosphere: The Myth of Fossil Fuels*. New York: Copernicus (2001).

Priest, Tyler, 'Hubbert's Peak: The Great Debate over the End of Oil' in *Historical Studies in the Natural Sciences* 44(1), pp.37–79 (February 2014).

WHAT IF ALIEN LIFE EXISTS ON EARTH?

Cleland, Carol & Copley, Shelley, 'The possibility of alternative microbial life on Earth' in *International Journal of Astrobiology* 4, pp.165–73 (2005).

Davies, Paul, et al., 'Signatures of a shadow biosphere' in *Astrobiology* 9(2), pp.241–9 (2009).

Davies, Paul, *The Eerie Silence: Are We Alone in the Universe?* New York: Allen Lane (2010).

WEIRD BECAME (PARTIALLY) TRUE: THE GAIA HYPOTHESIS

Lovelock, J. E., *Gaia: a new look at life on Earth*. New York: Oxford University Press (1979).

Ruse, Michael, *The Gaia Hypothesis: Science on a pagan planet*. Chicago: University of Chicago Press (2013).

Tyrrell, Toby, *On Gaia: a critical investigation of the relationship between life and Earth*. Princeton: Princeton University Press (2013).

WHAT IF WE'VE ALREADY FOUND EXTRATERRESTRIAL LIFE?

DiGregorio, Barry, *Mars: The Living Planet*. Berkeley, California: North Atlantic Books (1997).

Klein, Harold P., 'Did Viking Discover Life on Mars?' in *Origins of Life and Evolution of the Biosphere* 29(6), pp.625–31 (December 1999).

Levin, G. V., 'The Viking Labeled Release Experiment and Life on Mars' in *Proceedings of SPIE – The International Society for Optical Engineering*. San Diego, California (29 July–1 August 1997).

Levin, G. V. & Straat, P. A., 'Color and Feature Changes at Mars Viking Lander Site' in *Journal of Theoretical Biology* 75, pp.381–90 (1978).

Levin, G. V. & Straat, P. A., 'The Case for Extant Life on Mars and its Possible Detection by the Viking Labeled Release Experiment' in *Astrobiology* 16(10), pp.798–810 (2016).

Chapter Four: The Rise of the Psychedelic Ape

WHAT IF THE DINOSAURS DIED IN A NUCLEAR WAR?

Magee, M., *Who lies sleeping? The dinosaur heritage and the extinction of man*. Selwyn: AskWhy! Publications (1993).

McLoughlin, J. C., 'Evolutionary Bioparanoia' in *Animal Kingdom*, pp.24–30 (April/May 1984).

WHAT IF OUR ANCESTORS WERE AQUATIC APES?

Hardy, A., 'Was man more aquatic in the past?' in *New Scientist* 7, pp.642–5 (17 March 1960).

Kossy, D., *Strange Creations: Aberrant Ideas of Human Origins from Ancient Astronauts to Aquatic Apes*. Los Angeles: Feral House (2001).

Langdon, J. H., 'Umbrella hypotheses and parsimony in human evolution: a critique of the Aquatic Ape Hypothesis' in *Journal of Human Evolution* 33, pp.479–94 (1997).

Morgan, E., *The Descent of Woman*. New York: Stein and Day (1972).

WEIRD BECAME TRUE: THE OUT OF AFRICA THEORY

Dart, Raymond A., 'Australopithecus africanus: the man-ape of South Africa' in *Nature* 2884(115), pp.195–9 (7 February 1925).

Falk, Dean, *The Fossil Chronicles: How two controversial discoveries changed our view of human evolution*. Berkeley: University of California Press (2011).

Tobias, P. V., 'The life and times of Emeritus Professor Raymond A. Dart' in *South African Medical Journal* 67, pp.134–8 (25 January 1985).

WHAT IF WE'RE DESCENDED FROM A PIG–CHIMP HYBRID?

Critser, Greg, 'How a wild pig may uproot the tree of life' in *Quarterly Journal of the Los Angeles Review of Books* 7, pp.122–9. (Summer 2015).

McCarthy, Eugene M., *On the Origins of New Forms of Life: A New Theory*. Macroevolution.net (2008).

McCarthy, Eugene M., 'The Hybrid Hypothesis', Macroevolution.net. Retrieved from http://www.macroevolution.net/hybrid-hypothesis-contents. html.

Myers, P. Z., 'The MFAP Hypothesis for the origins of *Homo sapiens*', Pharyngula (2 July 2013). Retrieved from https://freethoughtblogs.com/ pharyngula/2013/07/02/ the-mfap-hypothesis-for-the-origins-of-homo-sapiens/

Prothero, Donald, 'Hogwash!', Skepticblog (4 December 2013). Retrieved from http://www.skepticblog.org/2013/12/04/hogwash/

WHAT IF HALLUCINOGENIC DRUGS MADE US HUMAN?

Clarke, D. B. & Doel, M. A., 'Mushrooms in post-traditional culture: apropos of a book by Terence McKenna' in *Journal for Cultural Research* 15(4), pp.389–408 (2011).

Huxtable, R. J., 'The mushrooming brain' in *Nature* 356, pp.635–6 (1992).

McKenna, T., *Food of the Gods: The Search for the Original Tree of Knowledge – A Radical History of Plants, Drugs, and Human Evolution*. New York: Bantam Books (1992).

Sagan, D., *Cosmic Apprentice*. Minneapolis: University of Minnesota Press (2013).

Self, W., 'Mushrooms Galore' in *Times Literary Supplement*, pp.7–8 (22 January 1993).

WEIRD BECAME TRUE: CAVE ART

Beltran, Antonio (ed.), *The Cave of Altamira*. New York: Harry M. Abrams (1999).

Curtis, Gregory, *The Cave Painters: Probing the Mysteries of the World's First Artists*. New York: Alfred A. Knopf (2006).

WHAT IF HUMANITY IS GETTING DUMBER?

Crabtree, G. R., 'Our fragile intellect', Parts 1 & 2 in *Trends in Genetics* 29(1), pp.1–5 (2013).

Geary, D. C., *The Origin of Mind: Evolution of brain, cognition, and general intelligence*. Washington, DC: American Psychological Association (2005).

Kalinka, A. T., Kelava, I. & Lewitus, E., 'Our robust intellect' in *Trends in Genetics* 29(3), pp.125–7 (2013).

McAuliffe, K., 'The incredible shrinking brain' in *Discover* 31(7), pp.54–9 (2010).

Mitchell, K. J., 'Genetic entropy and the human intellect' in *Trends in Genetics* 29(2), pp.59–60 (2013).

Chapter Five: Mushroom Gods and Phantom Time

WHAT IF ANCIENT HUMANS WERE DIRECTED BY HALLUCINATIONS?

Jaynes. J., *The Origin of Consciousness in the Breakdown of the Bicameral Mind*. Boston: Houghton Mifflin Company (1976).

Keen, S., 'The Lost Voices of the Gods: Reflections on the Dawn of Consciousness' in *Psychology Today* 11, pp.58–64, pp.66–7, pp.138–42, p.144 (1977).

Rowe, B., 'Retrospective: Julian Jaynes and The Origin of Consciousness in the Breakdown of the Bicameral Mind' in *The American Journal of Psychology* 125(3), pp.369–81 (2012).

Williams, G., 'What is it like to be nonconscious? A defense of Julian Jaynes' in *Phenomenology and the Cognitive Sciences* 10, pp.217–39 (2011).

WEIRD BECAME PLAUSIBLE: BEER BEFORE BREAD

Braidwood, Robert J., et al., 'Did man once live by beer alone?' in *American Anthropologist* 55, pp.515–26 (1953).

Hayden, Brian, Canuel, Neil & Sanse, Jennifer, 'What was Brewing in the Natufian? An Archaeological Assessment of Brewing Technology in the Epipaleolithic' in *Journal of Archaeological Method and Theory* 20, pp.102–50 (2013).

Katz, Solomon H. & Voigt, Mary M., 'Bread and Beer' in *Expedition* 28(2), pp.23–34 (1986).

WHAT IF HOMER WAS A WOMAN?

Beard, M., 'Why Homer Was (Not) a Woman: The Reception of the Authoress of the Odyssey' in *Samuel Butler, Victorian Against the Grain*. Toronto: University of Toronto Press (2007).

Butler, S., *The Authoress of the Odyssey, where and when she wrote, who she was, the use she made of the Iliad, and how the poem grew under her hands*. London: Longmans, Green (1897).

Dalby, A., *Rediscovering Homer: Inside the Origins of the Epic*. New York: W. W. Norton & Company (2006).

WHAT IF JESUS WAS A MUSHROOM?

Allegro, John M., *The Sacred Mushroom & the Cross*. London: Hodder and Stoughton (1970).

Brown, Judith Anne, *John Marco Allegro: The Maverick of the Dead Sea Scrolls*. Grand Rapids, Michigan: Wm. B. Eerdmans (2005).

King, John C., *A Christian View of the Mushroom Myth*. London: Hodder and Stoughton (1970).

WEIRD BECAME TRUE: ANCIENT TROY

Allen, Susan Heuck, *Finding the Walls of Troy: Frank Calvert and Heinrich Schliemann at Hisarlik*. Berkeley: University of California Press (1999).

Brackman, Arnold C., *The Dream of Troy*. New York: Mason & Lipscomb (1974).

Korfmann, Manfred, 'Was there a Trojan War?' in *Archaeology* 57(3), pp.36–8 (May/June 2004).

WHAT IF JESUS WAS JULIUS CAESAR?

Carotta, F., *Jesus was Caesar: On the Julian Origin of Christianity*. The Netherlands: Aspekt (2005).

Carotta, F., 'The gospels as diegetic transposition: A possible solution to the Aporia "Did Jesus Exist?"' (2007) Retrieved from http://carotta.de/subseite/texte/articula/Escorial_en.pdf.

Ehrman, B., *How Jesus Became God: The Exaltation of a Jewish Preacher from Galilee*. New York, NY: HarperOne (2014).

WHAT IF THE EARLY MIDDLE AGES NEVER HAPPENED?

Illig, H., *Das Erfundene Mittelalter: Die größte Zeitfälschung der Geschichte*. Munich: ECON (1996).

Illig, H., 'Anomalous Eras – Best Evidence: Best Theory', Toronto Conference (2005).

Niemitz, H-U, 'Did the Early Middle Ages Really Exist?' (1997) Retrieved from http://www.cl.cam.ac.uk/~mgk25/volatile/Niemitz-1997.pdf.

Scott, E., *A Guide to the Phantom Dark Age*. New York: Algora Publishing (2014).